TFET Integrated Circuits

Navneet Gupta • Adam Makosiej • Amara Amara
Andrei Vladimirescu • Costin Anghel

TFET Integrated Circuits

From Perspective Towards Reality

 Springer

Navneet Gupta
Institut Supérieur d'Électronique de Paris
Paris, France

Amara Amara
Institut Supérieur d'Électronique de Paris
Paris, France

Costin Anghel
Institut Supérieur d'Électronique de Paris
Paris, France

Adam Makosiej
Commissariat 'a l'Énergie Atomique et aux
Énergies Alternatives
Grenoble, France

Andrei Vladimirescu
Institut Supérieur d'Électronique de Paris
Paris, France

ISBN 978-3-030-55121-6   ISBN 978-3-030-55119-3   (eBook)
https://doi.org/10.1007/978-3-030-55119-3

This Springer imprint is published by the registered company Springer Nature Switzerland AG
The registered company address is: Gewerbestrasse 11, 6330 Cham, Switzerland

Contents

1 Introduction 1

2 State-of-the-Art TFET Devices 5
  2.1 TFET Introduction 5
  2.2 Silicon TFET Device TCAD and SPICE Models 8
  2.3 TFET Power and Delay Comparison with CMOS 11

3 SRAMs 17
  3.1 Introduction 17
  3.2 TFET SRAMs - State of the Art...................................... 18
    3.2.1 6T-TFET SRAM Design 18
    3.2.2 Other TFET SRAM Topologies.............................. 21
  3.3 Dual-Port SRAM Architecture 26
    3.3.1 Dual-Port TFET SRAM Cell................................. 27
    3.3.2 Micro-Architecture Design................................... 29
    3.3.3 Read Operation 29
    3.3.4 Single-Ended Sensing Methods 30
    3.3.5 Write Operation.. 30
    3.3.6 8T-TFET SRAM Stability and Performance 32
  3.4 3T-TFET Bitcell-Based TFET-CMOS Hybrid Memory 34
    3.4.1 Proposed 3T-TFET SRAM Cell.............................. 34
    3.4.2 Memory Architecture... 38
    3.4.3 Memory Layout... 39
    3.4.4 Energy Efficiency 40
    3.4.5 Read and Write Performance................................. 42
  3.5 Summary 43

4 Ultimate-D/SRAMs/CAMs.. 45
  4.1 Introduction 45
  4.2 The Ultimate-DRAM (uDRAM): TFET Negative-Differential-Resistance-Based 1T1C Refresh-Free DRAM... 46

	4.2.1	Write Operation	48
	4.2.2	Read Operation	49
	4.2.3	uDRAM Bitcell Implementation and Performance	51
4.3	uDRAM-Based 2T1C SRAM		52
4.4	uDRAM-Based 3T1C CAM		54
4.5	Summary		55

5 TFET *NDR* Flip-Flop 59
- 5.1   Introduction 59
- 5.2   State-of-the-Art TFET Flip-Flops 60
- 5.3   12T-TFET Master-Slave Flip-Flop (MSFF) Design 63
   - 5.3.1 Flip-Flop Operation 64
   - 5.3.2 Energy Efficiency 65
   - 5.3.3 Performance 66
- 5.4   Comparison and Results 68
   - 5.4.1 Setup Time (T_{Setup}) 69
   - 5.4.2 Clock-to-Output Propagation Delay (T_{CP2Q}) 69
   - 5.4.3 Minimum Clock Period Required $T_{Critical}$ 70
   - 5.4.4 Leakage 70
- 5.5   Summary 71

6 Content-Addressable Memories 73
- 6.1   Introduction 73
- 6.2   Hybrid TFET Reconfigurable CAM/SRAM Array Based on a
         9T-TFET Bitcell 75
   - 6.2.1 Write Operation (CAM and SRAM Modes) 76
   - 6.2.2 Read Operation (CAM and SRAM Modes) 77
   - 6.2.3 Implementation, Results, and Comparison 79
   - 6.2.4 Summary 82
- 6.3   ReConfigurable CAM Extension to CMOS 82
   - 6.3.1 ReCSAM Architecture 83
   - 6.3.2 Measured Results 86
   - 6.3.3 Summary 88
- 6.4   Associative Memory Architecture 90
   - 6.4.1 Associative Memory Architecture with CMOS CAM
            Bitcell 90
   - 6.4.2 Alternate TFET and CMOS Architectures 95
   - 6.4.3 Summary 100
- 6.5   Ultra-Low-Power TFET Ternary CAM (TCAM) 100
   - 6.5.1 TCAM Cell 100
   - 6.5.2 Summary 101

7 Sensing Techniques 105
- 7.1   Introduction 105
- 7.2   Charge-Injection-Based Single-Ended Imbalanced Sense
         Amplifier 106

7.3   Charge-Injection-Based Single-Ended Imbalanced Sense
      Amplifier with Open-Bitline... 108
7.4   TFET *NDR* Skewed Inverter-Based Sensing Method 110
      7.4.1   Sensing Method.. 111
      7.4.2   Summary 112
7.5   Adaptive Read Technique.. 113
      7.5.1   Adaptive Read Technique Operation 113
      7.5.2   Detailed Description of Functionality 114
      7.5.3   Alternative Design Options 119
      7.5.4   Summary and Applications.................................... 120

8 **Perspective** 123
   8.1   TFET-CMOS Hybrid Cores 123

References.. 127

Index... 135

List of Figures

Fig. 2.1   CMOS and TFET transistors 6
Fig. 2.2   Schematic representation of the energy-band diagrams for
           NTFET. OFF state: $V_D = 1$ V, $V_G = 0$ V-black curves; ON
           state: $V_D = 1$ V, $V_G = 1.6$ V-red curves 6
Fig. 2.3   TFET and MOSFET $I_D = f(V_G)$ 7
Fig. 2.4   Ambipolar effect: TFET conducts for both positive and
           negative V_G 7
Fig. 2.5   TFET I_D in reverse bias, $V_{DS} < 0$ V, $V_G = 0.25$–1.5 V 8
Fig. 2.6   Schematic representation of the simulated DG TFET
           structure with low-k spacer and high-k gate dielectrics [1] 9
Fig. 2.7   32 nm TFET and PTM MOSFET I_D (V_{GS}); V_{DS} step:
           0.25 V [©2016 IEEE] 10
Fig. 2.8   Output characteristics of the TFET for forward biasing;
           $I_D(V_{DS})$ [©2016 IEEE] 10
Fig. 2.9   Output characteristics of the TFET for reverse biasing;
           $I_D(V_{DS})$ step: 0.25 V [©2016 IEEE] 11
Fig. 2.10 Total Gate Capacitance of a PMOS (solid lines) and PTFET
           (dotted lines) as a function of gate voltage for different drain
           voltage values; V_{DS} step: 0.25 V [©2016 IEEE] 11
Fig. 2.11 Frequency vs. supply voltage for TFET and PTM [©2016
           IEEE] 12
Fig. 2.12 Power delay product (PDP) of the TFET and PTM [©2016
           IEEE] 12
Fig. 2.13 The result of the nine-stage RO simulation performed using
           TFETs at $V_{DD} = 1$ V [©2016 IEEE] 13
Fig. 2.14 Dynamic and static power vs. supply voltage for TFET and
           CMOS PTM [©2016 IEEE] 14
Fig. 2.15 P_{dyn}/P_{stat} of a nine stage TFET and CMOS RO 14

Fig. 2.16   TFET device symbols 15

Fig. 3.1   Typical bulk CMOS 6T-SRAM operation during *read*, (**a**)
            and *write* (**b**); the TFET 6T-SRAM operation during *read*
            (**c**) and *write* (**d**) [©2009 IEEE] 18
Fig. 3.2   3 × 3 bitcell array showing *HS* and *WD* problem 20
Fig. 3.3   3 × 3 TFET bitcell array showing *WD* problem 21
Fig. 3.4   8T-TFET bitcells [©2011 IEEE] 22
Fig. 3.5   7T-TFET bitcell [©2011 IEEE] 22
Fig. 3.6   Modified 6T-TFET bitcell [©2010 IEEE] 23
Fig. 3.7   5T-TFET-CMOS hybrid bitcell 23
Fig. 3.8   4T-*NDR* TFET bitcell 24
Fig. 3.9   8T-TFET dual-wordline bitcell [©2012 IEEE] 26
Fig. 3.10   Block diagram of proposed dual-port scratchpad [©2015
            IEEE] 27
Fig. 3.11   8T-TFET dual-port SRAM cell [©2015 IEEE] 28
Fig. 3.12   8T dual-port cell layout with two bitline connections. Top
            connections made with Metal-2 for BLR1&2, BLW1&2,
            VDD, GND (vertical) and Metal-3 for WL1 and WL2
            (horizontal) [©2015 IEEE] 28
Fig. 3.13   32 Kb Mux-8 dual-port memory architecture [©2015 IEEE] 29
Fig. 3.14   *Write* waveform with 0.9 V Supply [©2015 IEEE] 31
Fig. 3.15   Comparison of *write*-assist techniques [©2015 IEEE] 31
Fig. 3.16   Minimum wordline pulse width requirement vs. memory
            supply [©2015 IEEE] 32
Fig. 3.17   Static noise margins (*Read/Write*) vs. memory supply
            [©2015 IEEE] 33
Fig. 3.18   Leakage power vs. memory supply [©2015 IEEE] 33
Fig. 3.19   Proposed 3T-SRAM cell architectures [©2016 IEEE] 35
Fig. 3.20 I_D for DC sweep on node Qint at 0.6 V cell supply [©2016
            IEEE] 35
Fig. 3.21   TFET terminal voltages for different cell states [©2016
            IEEE] 36
Fig. 3.22   *Write* operation (Write-"0," Write-"1") [©2016 IEEE] 37
Fig. 3.23   *Read* operation waveform [©2016 IEEE] 38
Fig. 3.24   Bitcell and corresponding signal organization in mxn-SRAM
            cell array [©2016 IEEE] 38
Fig. 3.25   Proposed TFET/CMOS hybrid memory architecture [©2016
            IEEE] 39
Fig. 3.26   Cell array and layout [©2016 IEEE] 40
Fig. 3.27   *Read/Write* performance (WLPcrit) vs. cell supply voltage,
            including improvements with *read/write*-assist techniques
            [©2016 IEEE] 43

Fig. 4.1   TFET DRAM bitcell in retention, storing logical "0" and
            "1" [©2017 IEEE] 47

Fig. 4.2 TFET $I_D = f$ (reverse-biased V_{DS}) characteristics and
capacitor leakage [©2017 IEEE] 47
Fig. 4.3 Signals during *write* [©2017 IEEE] 48
Fig. 4.4 *Write* waveforms [©2017 IEEE] 49
Fig. 4.5 Signals during *read*, (**a**) selected cell and (**b**)
partially-selected cells due to precharged BLs [©2017 IEEE] 50
Fig. 4.6 *Read* waveforms [©2017 IEEE] 50
Fig. 4.7 2 × 2 Bitcell array organization [©2017 IEEE] 51
Fig. 4.8 uDRAM bitcell layout [©2017 IEEE] 51
Fig. 4.9 2T1C uSRAM bitcell (retention bias voltages) [©2017
IEEE] 52
Fig. 4.10 *Read* waveforms showing two RBLs (read "0" and read "1")
with active low WL. Internal cell nodes retain data during
read [©2017 IEEE] 53
Fig. 4.11 2T1C uSRAM bitcell array organization [©2017 IEEE] 53
Fig. 4.12 uSRAM bitcell layout 54
Fig. 4.13 3T1C uCAM bitcell 54
Fig. 4.14 uCAM *read*: hit and miss conditions 55
Fig. 4.15 uCAM 2x2 bitcell array 56
Fig. 4.16 uCAM bitcell layout 56

Fig. 5.1 TFET transmission-gate flip-flop [©2013 IEEE] 60
Fig. 5.2 TFET master-slave flip-flop [©2013 IEEE] 61
Fig. 5.3 TFET semi-dynamic flip-flop [©2013 IEEE] 61
Fig. 5.4 TFET sense amplifier-based latch [©2013 IEEE] 62
Fig. 5.5 TFET sense amplifier-based flip-flop [©2013 IEEE] 62
Fig. 5.6 TFET pseudo-static flip-flop [©2013 IEEE] 63
Fig. 5.7 Proposed 12T (14T with O/P driver-2) TFET-FF design
[©2016 IEEE] 64
Fig. 5.8 Flip-flop voltage waveforms, (**a**) Input data and clock, (**b**)
inverted output, and (**c**) FF internal nodes 65
Fig. 5.9 12T-TFET FF comparison with LP-CMOS and FinFET
FF's: leakage power vs. supply voltage [©2016 IEEE] 66
Fig. 5.10 12T-TFET FF comparison with LP-CMOS and FinFET
FF's: dynamic power consumption vs. supply voltage
[©2016 IEEE] 66
Fig. 5.11 12T-TFET FF comparison with LP-CMOS and FinFET
FF's: T_{Setup} vs. supply voltage [©2016 IEEE] 67
Fig. 5.12 12T-TFET FF comparison with LP-CMOS and FinFET
FF's: T_{CP2Q} vs. supply voltage [©2016 IEEE] 67
Fig. 5.13 12T-TFET FF comparison with LP-CMOS and FinFET
FF's: maximum operating frequency vs. supply voltage
[©2016 IEEE] 68
Fig. 5.14 Comparison—T_{Setup} for different flip-flop designs [©2016
IEEE] 69

Fig. 5.15   Comparison—T_{CP2Q} for different designs [©2016 IEEE]   70
Fig. 5.16   Comparison—$T_{Critical}$ for different designs [©2016 IEEE]   71

Fig. 6.1   9T-TFET CAM cell [©2016 IEEE]   75
Fig. 6.2   ReCSAM architecture [©2016 IEEE]   76
Fig. 6.3   CAM *write* operation [©2016 IEEE]   77
Fig. 6.4   CAM *search* operation, word[0]-CAM hit (BLL0), and
            word[1]-worst-case CAM miss (BLL1) [©2016 IEEE]   78
Fig. 6.5   Cell array and dual-bitcell layouts; VDD, BLR, and BLL in
            Metal-2; GND, WL1, WL2, and RBL in Metal-3 [©2016
            IEEE]   79
Fig. 6.6 WLP_{MIN} for *read/write* operations for CAM and SRAM
            modes vs. supply voltage [©2016 IEEE]   80
Fig. 6.7   Memory organization [©2017 IEEE]   83
Fig. 6.8   Single-ended imbalanced sense amplifier, precharged to "0"
            [©2017 IEEE]   84
Fig. 6.9   Simulation waveforms of single-ended *read* during CAM
            search [©2017 IEEE]   85
Fig. 6.10   Chip photo and test-macro layout [©2017 IEEE]   86
Fig. 6.11   *Read/write* performance [©2017 IEEE]   87
Fig. 6.12   *Read* speed with/without SA imbalance tuning [©2017
            IEEE]   87
Fig. 6.13   VDDmin distribution for ten chips and bitcell leakage. (**a**)
            VDDmin distribution. (**b**) Bitcell Leakage [©2017 IEEE]   88
Fig. 6.14   Assist techniques analysis, (**a**) WL boosting, (**b**) negative
            bitline [©2017 IEEE]   88
Fig. 6.15   Associative memory architecture   91
Fig. 6.16   6T-CAM bitcell   91
Fig. 6.17   Winner-take-all implementation for two columns   92
Fig. 6.18   Winner-take-all logic for one memory bank; one WTA per
            column   94
Fig. 6.19   *Search* operation waveform, simulation condition
            (mismatch): Col[0]-4 bit, Col[1]-3 bit, Col[2]-1 bit   94
Fig. 6.20   *Write* operation with bitcell ground at $VDD/2$ 95
Fig. 6.21   TFET associative memory architecture   96
Fig. 6.22   8T-TFET associative memory compatible bitcell   96
Fig. 6.23   10T-CMOS CAM cell   98
Fig. 6.24   Bitcell array of 10T-NOR CAM-based associative memory   98
Fig. 6.25   Overall memory organization with WTA logic on *match* lines   98
Fig. 6.26   Current-mode WTA circuit   99
Fig. 6.27   Current-mode WTA-based comparison logic for associative
            memories   99
Fig. 6.28   Associative memory test-chip photo and block layout
            description   100
Fig. 6.29   Hybrid TFET/CMOS 7T-TCAM bitcell [©2016 IEEE]   101

Fig. 6.30   *Read* waveform for hit and miss conditions [©2016 IEEE] 102
Fig. 6.31   *Write* waveform for writing X (10), 0 (00), and 1 (11)
            [©2016 IEEE] 102
Fig. 6.32   Memory array organization [©2016 IEEE] 102
Fig. 6.33   TCAM dual-bitcell layout [©2016 IEEE] 103

Fig. 7.1   Imbalanced sense amplifier circuit diagram [©2015 IEEE] 107
Fig. 7.2   Waveform: *Read* "0" using imbalanced sense amplifier
            [©2015 IEEE] 107
Fig. 7.3   Waveform: *Read* "1" using imbalanced sense amplifier
            [©2015 IEEE] 108
Fig. 7.4   DPSRAM memory: open-bitlines architecture [©2015
            IEEE] 109
Fig. 7.5   Circuit diagram—(**a**) Sense amplifier, (**b**) Memory output
            driver [©2015 IEEE] 109
Fig. 7.6   Waveforms for simultaneous *read* on both ports (see
            Fig. 7.4, (**a**) Port-1 (read "0"), (**b**) Port-2 (read "1") [©2015
            IEEE] 110
Fig. 7.7   Proposed sense amplifier *Read* circuit [©2016 IEEE] 111
Fig. 7.8   DC characteristics with different sizing of M_0 with widths
            of 200 nm, 400 nm, 600 nm, and 800 nm [©2016 IEEE] 111
Fig. 7.9   Different trip points for sensing ($5fF$ load); M_0
            sizing—400 nm blue; 600 nm pink; 800 nm orange [©2016
            IEEE] 112
Fig. 7.10   *Read* delay vs. supply voltage for TFET-CMOS SA with a
            $5fF$ load [©2016 IEEE] 113
Fig. 7.11   Schematic: proposed sense amplifier 114
Fig. 7.12   Schematic: self-adaptive sense amplifier 115
Fig. 7.13   Memory architecture with $RdOK_{MEM}$ generation 116
Fig. 7.14   Waveforms: reliable *Read* operation 117
Fig. 7.15   Waveforms: unreliable *Read* operation 117
Fig. 7.16   Memory architecture with self-tuning 118
Fig. 7.17   Self-adapt memory architecture 118
Fig. 7.18   Waveforms: adaptive *Read* 119
Fig. 7.19   Schematic: adaptive single-ended sense amplifier 120
Fig. 7.20   Adaptive memory test-chip photo and layout details 121

List of Tables

Table 3.1   Comparison - power and area [©2016 IEEE] 41

Table 4.1   uDRAM, uSRAM, and uCAM comparison with
state-of-the-art CMOS RAM 57

Table 5.1   Transistor count for various flip-flop designs [©2016 IEEE] 68

Table 6.1   Signal voltages during CAM-/SRAM-mode *write* for
selected and *HS* cells (writing) 77

Table 6.2   Signal voltages during CAM- and SRAM-mode *read*
[©2016 IEEE] 78

Table 6.3   TFET CAM/SRAM speed and power comparison [©2016
IEEE] 81

Table 6.4   CMOS CAM/SRAM comparison with state-of-the-art
[©2017 IEEE] 89

Table 6.5   Signal voltages during *Search* 93

Table 6.6   Voltages during *write* for selected and *HS* cells (writing "1") 95

Chapter 1
Introduction

Previous trends in System-on-Chip (SoC) design were focused on improving the performance of the system without giving significant consideration to power consumption. Complementary-Metal-Oxide-Semiconductor (CMOS) is the widely accepted technology for designing SoCs for over three decades. Performance improvements of CMOS systems came mostly through technology scaling and when speed could not be increased any further, the growing number of transistors per chip, which followed Moore's law, led to multi-core processor chips. Technology scaling significantly improved the system performance and allowed to increase complexity of systems in cost-effective ways. However, power consumption became the major constraint in design specifications because of increased leakage with every new technology node.

The development of energy-efficient systems is becoming ever more important with widely increasing use of battery-operated systems for applications such as Internet-of-Things (IoT) and Wireless-Sensor-Nodes (WSN). These systems have an ever increasing need of energy efficiency for longer battery life time while maintaining performance to meet the application requirements. Unlike previous trends in computation focused on increasing the performance at the cost of energy and area, for applications such as WSN and IoT, system cost, energy efficiency, and performance are equally important today. To be successful in the market these new systems should be designed with low power and low cost in mind. IoT and WSN applications can have periods of low activity and long standby times with leakage becoming a critical concern for battery life time in SoCs designed for these applications. Moreover, energy efficiency is becoming increasingly important because of the ever increasing power density and costly heat sinks required in compute-intensive systems.

As ultra-low power consumption is a crucial feature of IoT and WSN, they require major advances in the field of semiconductor devices, integrated circuits, and system architectures [2], which go well beyond the improvements predicted by Moore's law. Indeed, energy/power consumption reductions coming from CMOS

© Springer Nature Switzerland AG 2021
N. Gupta et al., *TFET Integrated Circuits*,
https://doi.org/10.1007/978-3-030-55119-3_1

scaling are now planned to be in the order of 3X in the next decade [3]. In other words, the traditional reliance on CMOS technology scaling is definitely an inadequate approach from the perspective of the IoT evolution, and much greater reductions in consumption need to come from combined technology, circuit and architectural breakthroughs [2]. Also, since IoT and WSNs need to be generally inexpensive (in the dollar range), compatibility with CMOS technology is a must.

Due to the above-mentioned factors efforts were made at both system and circuit level to optimize both dynamic and static power consumption using various techniques such as dynamic-voltage-frequency-scaling (DVFS), power gating, reconfigurable computing with shared resources [4–6], high-throughput and small area embedded-DRAMs [7], and sacrificing area to reduce leakage while maintaining sufficient performance [8]. DVFS is a particularly important technique in the IoT world due to the operating-mode-dependent frequency requirement, which ranges from kHz to MHz. Digital systems consist mainly of logic gates, flip-flops, and memories; therefore, power optimization of each of these components is an important design aspect. One approach proposed for logic [9] is to use power-gated standard cells for reducing standby leakage.

For IoT applications, optimizing memory power is of critical concern as more than 50% and 90% of total power consumption and leakage in standby, respectively, is in memory [10, 11]. In [11] the authors reported processors with 39% and 51% of total standby power dissipated in instruction memory and data memory, respectively. Moreover, in processors significant die area is consumed by cache memories; as an example a 37.5 MB cache consumes more than 25% of the die area [7]. Thus, it is of the utmost importance to optimize memory area and leakage power consumption simultaneously.

In memories, leakage reduction at the cost of bitcell area increase is an inefficient optimization method because a larger size bitcell results in significant area penalty at array level. Therefore, lowering the voltage of operation is a widely used leakage reduction technique in memories. However, this causes bitcell stability issues at low voltage forcing the designers to use either assist techniques or bigger bitcells. Moreover, the bigger footprint and power budget of SRAMs are forcing designers to limit the total on-chip memory size. These different constraints make SoC memory design at advanced technology nodes very challenging.

The objective of this book is to analyze the potential of devices with very-low leakage such as Tunnel-Field-Effect Transistors (TFETs) for low-power circuits and systems applications; these devices have already been proposed as replacement of standard CMOS for overcoming its limitations. The main focus of the researchers is to design SRAMs, with compact cells, low power, high speed, and good stability. The optimizations are done at technology, cell, and architecture levels. But low-power compact cells with high speed are still missing, especially those with ultra-low leakage.

The TFET operates by quantum-tunneling effect, which is different from the MOSFET working principle. Therefore, TFETs do not suffer from the subthreshold slope limitation of 60 mV/decade as MOSFETs [12, 13]. While optimized TFETs can provide leakage currents on the order of fA/μm [14], one major concern is the ON current, which is lower than CMOS. However, a few recent reports demonstrated fabricated TFETs with drain currents up to $760\,\mu A/\mu m$ [15] and subthreshold slopes as low as $30\,mV/decade$ [16].

These results confirm the TFET's potential for successful utilization in low-power/standby power applications and encouraged research on TFET circuits. Our goal is aligned with the reports present in literature and our efforts aim to explore technology and circuit architectures including memories, with reduced power consumption, especially static power, low area, and high speed of operation.

From the above considerations TFETs fit very well the chip power/cost profile that is imposed by IoT and WSN applications. However, TFETs are significantly slower than CMOS [17], in spite of device improvements demonstrated recently [1, 14, 15]. Although the recent work on TFETs has targeted the same device-level goals as CMOS transistors along the International Technology Roadmap for Semiconductors (ITRS) [3], *our vision is that TFETs will not replace CMOS, rather they will complement it.* Instead of simplistically trying to match and replace CMOS, our vision permits to retain the benefits of a mature technology such as CMOS and the related existing ecosystem, while leveraging the unique properties and the true potential of TFETs, which have a far lower leakage but lower speed.

The focus of this text is the exploration of new solutions for low-power circuits using TFETs while at the same time addressing critical issues at circuit and architecture level. In our research work we analyzed and proposed topologies and architectures for various kind of circuits, including SRAMs, CAMs, TCAMs, DRAMs, flip-flops, latches, and sense amplifiers for sub-32 nm technologies.

The book is organized as follows. Chapter 2 introduces the TFET device and its characteristics; the TCAD device used in developing the circuits in the following chapters is described along with its compact model for simulation and its performance evaluation. Chapter 3 targets the investigation of Silicon-TFET circuits, compatible with CMOS, for Low STandby Power (LSTP) applications with ultra-low leakage for long battery life and/or energy harvesting. We analyze the architecture-level issues in known TFET SRAM designs, propose new TFET cells followed by developing hybrid architectures with CMOS to optimize speed, power, and area.

In Chap. 4 architectures/circuits are explored using a unique property of TFETs, *Negative Differential Resistance (NDR)*, and capacitor leakage. Novel TFET DRAM, SRAM, and CAM cells are proposed.

In Chap. 5 the exploration of TFET architectures/circuits using *NDR* is extended to flip-flops. A novel flip-flop design with low voltage, low power, and high-speed operation is proposed and compared with the state of the art.

In Chap. 6 the circuit architectures are extended to full integration in a CMOS platform to make them useful for existing technologies and products. Different CMOS architectures extended from TFET research and new architectures applicable to both TFET and CMOS are proposed. Circuits such as reconfigurable-CAMs, adaptive *read* for memories and approximate-search CAMs are proposed and designed to be implemented in a 28 nm FDSOI-CMOS process. The circuit and layout design is performed using sub-32 nm TFETs [14] and 28 nm FDSOI CMOS technology [18]. These two types of devices are compatible for fabrication in a single FDSOI-CMOS process.

In Chap. 7 single-ended sensing techniques and adaptive *read* techniques are explored for SRAMs and original SRAM, DRAM, and CAM architectures presented in Chaps. 3, 4, and 6, respectively.

Finally, Chap. 8 offers a perspective of innovative applications of TFET's unique features to hybrid systems with tighter cointegration of TFET and CMOS.

Chapter 2
State-of-the-Art TFET Devices

2.1 TFET Introduction

TFETs are p-i-n gated junctions that operate in reverse regime. Figure 2.1 shows a conceptual TFET structure compared to a CMOS transistor. For an n-type (NTFET), p+ (n+) doping is used for the source (drain), while for a p-type (PTFET) n+ (p+) doping is used for the source (drain), the doping being reversed between the source and the drain as opposed to a MOSFET where drain and source have identical doping.

There are two types of TFETs, homo- and hetero-junction [15, 16, 19, 20], depending on the type of semiconductor of the p-i-n junctions. The former is implemented in Silicon, while the latter uses different III-V materials such as InAs or GaSb-InAs. In this text we are addressing Si TFETs due to their potential of being integrated with CMOS on the same substrate.

The operation of the TFET is based on band-to-band tunneling (BTBT) consisting in the modification of the position of the band gap of the intrinsic region of the device relative to the energy levels of the source and drain, see Fig. 2.2 [12, 13]. In the ON state a positive voltage is applied on the gate of an NTFET leading to sufficient narrowing of the band gap such that tunneling can occur. If the gate bias is low, close to 0 V, the band gap of the channel blocks the tunneling, corresponding to the OFF state in Fig. 2.2. The n+ drain is always biased with positive voltage, $V_{DS} > 0$ V, to ensure the operation in the reverse regime of the p-i-n diode; for the TFET this biasing represents the forward operation region. This polarization guarantees extremely low I_{OFF} currents when the device is OFF with $V_G = 0$ V.

TFET $I_D = f(V_G)$ typical characteristics are plotted in Fig. 2.3 along with those of a FDSOI MOSFET. It can be seen that the TFET has a lower I_{ON} compared to any MOSFET; on the flip side, however, the TFET operating by quantum-tunneling does not suffer from the subthreshold slope limitation as does a MOSFET, [12, 13, 20], and therefore, has a very low OFF current, I_{OFF}.

© Springer Nature Switzerland AG 2021
N. Gupta et al., *TFET Integrated Circuits*,
https://doi.org/10.1007/978-3-030-55119-3_2

Fig. 2.1 CMOS and TFET transistors

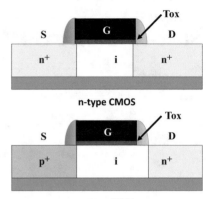

Fig. 2.2 Schematic representation of the energy-band diagrams for NTFET. OFF state: $V_D = 1$ V, $V_G = 0$ V-black curves; ON state: $V_D = 1$ V, $V_G = 1.6$ V-red curves

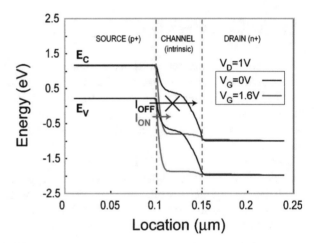

Experimental results have shown slopes less than 50 mV/decade and 10^{-14} A I_{OFF} currents at room temperature [16]. Regardless of these encouraging results, TFETs suffer from a low I_{ON}, well below the conduction current of CMOS.

The TFET is an ambipolar device, i.e., it conducts both for positive and negative V_{GS} with the BTBT occurring at the metallurgical source-channel junction or at the metallurgical drain-channel junction respectively, see Fig. 2.4. When positive or negative voltage is applied to the gate the bands in the channel region are pushed down at the source edge or up at the drain edge, respectively. This is highlighted in Fig. 2.2 for positive V_G when electrons can tunnel from the valence band of the source at the channel interface to the conduction band in the channel (drain). The ambipolar behavior is not desirable in circuit design.

The device used in this text, Structure 1 in Fig. 2.4, displays reduced ambipolar behavior having under-lapped gate and drain by 30 nm [1, 14].

Another unique feature of the TFET is its I_D characteristic in reverse bias, $V_{DS} < 0$ V and $V_{DS} > 0$ V for an NTFET and PTFET, respectively. As shown in Fig. 2.5

Fig. 2.3 TFET and MOSFET $I_D = f(V_G)$

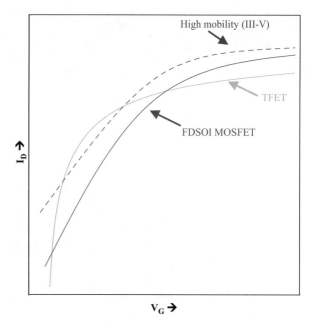

Fig. 2.4 Ambipolar effect: TFET conducts for both positive and negative V_G

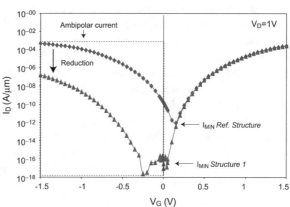

for very small absolute values of V_{DS} below 0 there is still current flowing due to BTBT, which only stops when V_{DS} gets below -100 to -200 mV. The conduction mechanism changes from drift to diffusion for $|V_{DS}| >= 0.6$ V corresponding to the turn-on voltage of the diode. The transition between the two conduction mechanisms creates a *Negative Differential Resistance* (*NDR*) region that can have interesting applications in circuit design as described in the following chapters.

Fig. 2.5 TFET I_D in reverse
bias, $V_{DS} < 0$ V,
$V_G = 0.25$–1.5 V

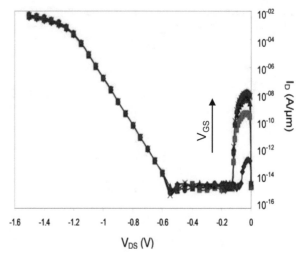

2.2 Silicon TFET Device TCAD and SPICE Models

A Silicon TFET device structure was created using TCAD [1] based on $I - V$ characteristics of measured devices and improved based on the authors' research for increased I_{ON} and reduced ambipolar behavior [14]. Our TFET structure shown in Fig. 2.6 is using Low-k (SiO2) spacers and a High-k (HfO_2) gate dielectric [14, 19], with the following dimensions: the gate and the spacers lengths are 30 nm each, the gate dielectric physical thickness is 3 nm, whereas the Silicon film (tSi) is 4 nm. The gate metal work-function is 4.4 eV.

The $I - V$ characteristics of the TFET are obtained from TCAD simulations, see Figs. 2.7, 2.8, and 2.9. The non-local BTBT and the bandgap-narrowing models were used in Silvaco Atlas (version 5.15.32.R) for the device simulation. The analytical model proposed by Niquet et al. [21] was used to estimate the silicon bandgap widening as a function of the film thickness.

The TCAD simulations were calibrated with respect to data presented in the literature [22, 23]. Figure 2.7 presents the $I_D(V_{GS})$ characteristics of a p-type TFET [24] in comparison to those of a 32 nm Predictive Technology Model (PTM) [25] pMOSFET model for V_{DS} varying from -0.25 V to -1.0 V. The TFET ON current I_{ON} is reduced when compared to that of the MOSFET due to the high resistance of the tunneling barrier. The essential TFET advantages over CMOS are: (1) the very low OFF-state current I_{OFF} and (2) the steep subthreshold slope S.

Another important parameter for both device types is the turn-on voltage. For a TFET this is V_{OFF} defined as the value of $I_D(V_{GS})$ where I_D bottoms out, as shown in Fig. 2.7. For a MOSFET the corresponding parameter is the threshold voltage (V_T) defined more arbitrarily as the value where the current reaches 100 nA.

Our devices are designed such as to obtain a V_{OFF} voltage of around 150 mV (Fig. 2.7) in contrast to other reports where V_{OFF} is typically very close to 0 V in

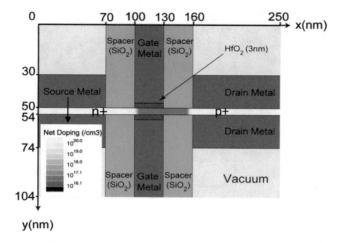

Fig. 2.6 Schematic representation of the simulated DG TFET structure with low-k spacer and high-k gate dielectrics [1]

order to maximize the dynamic performance. Our choice is based on the fact that $V_{OFF} > 0$ V guarantees low I_{OFF} currents also in the presence of the unavoidable random parameter variations. With a V_{OFF} of 0 V, even a slight variation of the TFET characteristics would lead to an important I_{OFF} increase in the device due to its steep slope around V_{OFF}. It is worth noting that the sources of TFET variability are different when compared to those of a MOSFET. For TFETs the tunneling barrier at the source-channel junction and source-gate overlap are the major factors for current variations. Few reports in literature [26, 27] show the analysis of process-induced variations and their modelling in TFETs.

To highlight the impact of process variation, let us consider a small-size device of $L = 30$ nm length and $W = 100$ nm width, and assume that the TFET V_{OFF} variability is similar to that of a bulk MOSFET V_T. The value of the Pelgrom coefficient (A_{VT}) [28] for a CMOS LP 32 nm process in the equation defining the threshold voltage standard deviation σ_{V_T} is approximately 2.5 mV μm. Applying the Pelgrom model for the standard deviation of V_{OFF}, Eq. (2.1), results for the device of the aforementioned size in a V_{OFF} variation of 45.6 mV and 136.9 mV for 1σ and 3σ probabilities, respectively. As a consequence, if V_{OFF} is located close to 0 V for a sub-60 mV/decade subthreshold-swing device like the TFET, the leakage current increase caused by random variations is over three orders of magnitude for a 3σ distribution of V_{OFF}. High V_{OFF} values of 150–170 mV as obtained on our devices guarantee low I_{OFF} variability if fabricated in a process with high A_{VT}.

$$\sigma_{V_{OFF}} = \frac{A_{VT}}{\sqrt{W * L}} \qquad (2.1)$$

The forward and reverse $I_D(V_{DS})$ characteristics of a PTFET are plotted in Figs. 2.8 and 2.9, respectively. The operation of TFET circuits is negatively affected

Fig. 2.7 32 nm TFET and PTM MOSFET I_D (V_{GS}); V_{DS} step: 0.25 V [©2016 IEEE]

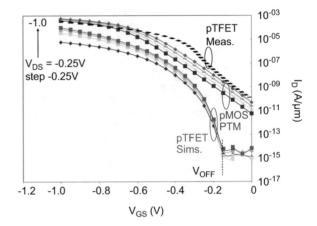

Fig. 2.8 Output characteristics of the TFET for forward biasing; $I_D(V_{DS})$ [©2016 IEEE]

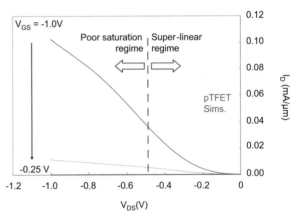

by the fact that the forward output characteristics I_D (V_{DS}) of the TFET does not saturate as shown in Fig. 2.8. On the simulated reverse output characteristics of the PTFET plotted in Fig. 2.9 it can be observed that similar to the NTFET the current is important for two biasing conditions: (1) at low positive (negative for an NTFET) V_{DS}, when the current is dominated by BTBT, and (2) at high positive (negative for an NTFET) V_{DS}, when the turn-on of the p-i-n junction occurs as it is now biased in forward mode. In the latter bias condition the gate has little control on the device current, see Fig. 2.9. In literature, this behavior is called *unidirectional* as the gate controls the characteristics of TFET only in the forward regime, Figs. 2.7 and 2.8, and not in the reverse regime, Fig. 2.9. Therefore, TFETs should not be biased in reverse with high negative V_{DS} for NTFETs (positive V_{DS} for PTFET) to avoid high-leakage currents, as can be seen in Fig. 2.9.

The total gate capacitance of a PTFET in comparison with that of a pMOSFET is plotted in Fig. 2.10. It can be noticed that the capacitance of a PTFET diminishes, for a given V_{GS} value, as the drain voltage $|V_{DS}|$ is increased when conducting in

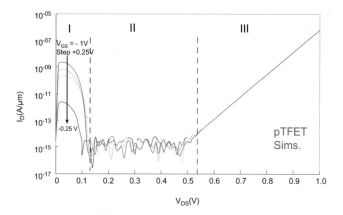

Fig. 2.9 Output characteristics of the TFET for reverse biasing; $I_D(V_{DS})$ step: 0.25 V [©2016 IEEE]

Fig. 2.10 Total Gate Capacitance of a PMOS (solid lines) and PTFET (dotted lines) as a function of gate voltage for different drain voltage values; V_{DS} step: 0.25 V [©2016 IEEE]

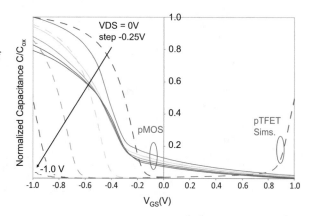

the forward region ($V_{GS} < 0$). This behavior is in contrast to MOSFETs where this reduction is not present.

For SPICE simulation of the circuits described in this book, both P and NTFETs were modeled using look-up tables. Both DC and capacitance characteristics were implemented as $I_D(V_{GS}, V_{DS})$, $C_{GS}(V_{GS}, V_{DS})$, and $C_{GD}(V_{GS}, V_{DS})$ tables obtained from TCAD simulations of Si-TFETs.

2.3   TFET Power and Delay Comparison with CMOS

In order to benchmark the TFET dynamic performance versus CMOS, a nine-stage Ring Oscillator (RO) was simulated using our 30 nm TFET and 32 nm LP PTM CMOS devices. To correctly compare the two implementations, the CMOS and

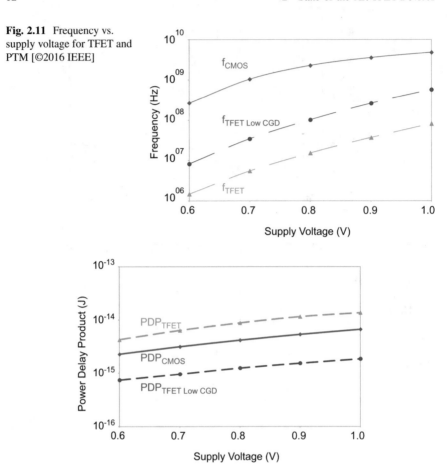

Fig. 2.11 Frequency vs. supply voltage for TFET and PTM [©2016 IEEE]

Fig. 2.12 Power delay product (PDP) of the TFET and PTM [©2016 IEEE]

TFET RO inverters were sized to obtain a similar total area while maintaining optimum-strength ratios, i.e. a balanced inverter voltage-transfer characteristic. For the CMOS RO, NMOS and PMOS are sized (W/L) at 130/30 nm and 170/30 nm, respectively. The sizes are decided on the basis of PTM 32 nm PMOS and NMOS currents in simulation. For the TFET RO, both PTFET and NTFET are set to 150/30 nm, since the current in PTFET and NTFET is almost equal. As depicted in Fig. 2.11 the TFET RO frequency is significantly lower than the one of the CMOS RO. The frequency ratio of the CMOS RO with respect to that of the TFET RO varies approximately from 2 to 5 orders of magnitude for operating voltages from 0.6 V to 1 V. The Power Delay Product (PDP) analysis in Fig. 2.12 reveals that the TFET RO PDP is higher than that of the CMOS RO by a factor of approximately 1.8–2 in the given voltage range.

The lower operating frequency of the TFET RO can be largely attributed to the lower TFET current drivability that leads to a lower average RO current. An

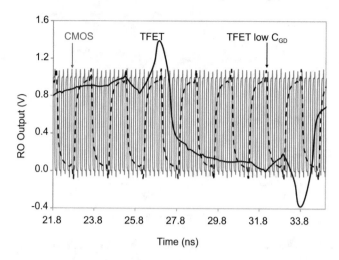

Fig. 2.13 The result of the nine-stage RO simulation performed using TFETs at $V_{DD} = 1$ V [©2016 IEEE]

additional factor contributing to the higher TFET PDP is linked to its larger Miller capacitance than that of CMOS [29], which should be as low as possible for high-speed operation of the RO. The high value of the Miller capacitance at each RO node lengthens the switching time and leads to high under- and overshoots in the output voltage, see Fig. 2.13. The impact of the Miller capacitance effect is demonstrated by simulating the TFET RO also with a fixed gate to drain capacitance (C_{GD}) of 0.114 fF/μm, which is the minimum C_{GD} value extracted from TCAD simulations of our device. By imposing a fixed low value of C_{GD} the effect of Miller capacitance is minimized. Figure 2.11 shows that with a fixed C_{GD} the TFET RO frequency is 8x higher than that of the actual TFET RO. The Miller capacitance impact on PDP can be seen in Fig. 2.12 where it can be observed that TFETs with low Miller capacitance outperform CMOS. Figure 2.13 compares waveforms for TFET ROs with variable $f(V_{DS})$ and fixed low C_{GD} with CMOS RO; the under- and overshoots for the TFET RO with fixed C_{GD} are 3× smaller and the RO frequency is 8× higher. In summary, we can conclude that the TFET lower dynamic performance is linked to three key properties of this device: (1) high Miller capacitance value, (2) low I_{ON}, and (3) strong I_{ON} dependency on V_{DS} (non-saturating), see Fig. 2.8. This leads to low frequency operation because of the difficulty in fully charging and discharging the large capacitances of RO nodes.

Reports in literature show hetero-junction TFETs having higher performance. However, designing reliable hetero-junction TFETs is much more difficult due to process immaturity in comparison to Si-TFETs, which can be implemented using a process similar to Si-CMOS. This work focuses on the design of Si-TFET integrated circuits for applications where these circuits outperform CMOS and the combination of the two achieves both performance and power savings on the same chip.

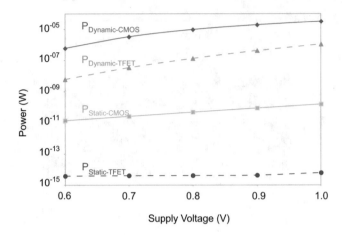

Fig. 2.14 Dynamic and static power vs. supply voltage for TFET and CMOS PTM [©2016 IEEE]

Fig. 2.15 P_{dyn}/P_{stat} of a nine stage TFET and CMOS RO

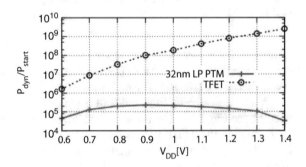

Despite a lower dynamic power efficiency, the TFET advantage over CMOS becomes clear by considering static operation in terms of the RO short-circuit current. Figure 2.14 depicts the dynamic (*Pdyn*) and static (*Pstat*) power of the CMOS PTM and TFET ROs plotted against V_{DD}. The difference in *Pdyn* between the two is consistent with the PDP behavior shown in Fig. 2.12 and as shown in Fig. 2.15 the ratio of *Pdyn* / *Pstat* varies from 2 to 5 orders of magnitude for V_{DD} from 0.6 V to 1.4 V. However, the *Pstat* ratio is in favor of the TFET significantly with 4 to over 6 orders of magnitude improvement over CMOS for V_{DD} between 0.6 V and 1.4 V, respectively.

Another TFET advantage is that its *Pstat* remains almost constant with the increase in supply voltage due to the fact that in a TFET I_D shows a very weak dependence on V_{DD} for $|V_{GS}| < |V_{OFF}|$ and remains in the fA range as depicted in Fig. 2.7. By comparison, in CMOS *Pstat* increases by approximately two orders of magnitude with the increase in supply voltage requiring lowering it in CMOS designs in order to reduce the leakage contribution to the overall power consumption. This is not needed for TFETs, due to very low *Pstat* dependence on

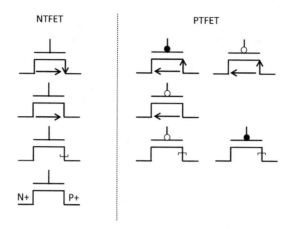

Fig. 2.16 TFET device symbols

V_{DD}, demonstrating its superiority in terms of power efficiency and its attractiveness over CMOS as long as very high operation frequency is not required (LP and LSTP designs).

This observation is also important for SRAM designs. As TFET leakage is in the fA range even for supply voltages of 1 V and above, it is no longer necessary to scale down the supply voltage to reduce SRAM leakage. As it will be seen, maintaining high V_{DD} in TFET SRAM designs allows for the compensation of their lower dynamic performance without increasing leakage (*Pstat*), which is not applicable to present CMOS designs.

In circuit schematics in this book more than one symbol is used for the TFET since there is no agreed-upon unique representation. Schematics representing our own designs use a symbol similar to that of a MOSFET with an arrow added indicating the source and the drain as the flow of the current is *unidirectional* unlike for MOSFETs; these symbols are shown in Fig. 2.16 for N and PTFET. Schematics taken from literature when reviewing previous art may use different symbols; however, the text will always clarify whether the devices are TFETs or MOSFETs.

Chapter 3
SRAMs

3.1 Introduction

The TFET emerges as one of the promising alternatives to CMOS to design ultra-low power memories due to very-low leakage current [4, 5, 14, 26, 30–33]. In literature, reports on optimizing TFET circuits are mainly focused on SRAM designs with the aim to reduce leakage [26, 30–32].

The key challenge in designing TFET-based SRAMs or CAMs is linked to the different-than-CMOS TFET characteristics, i.e., unidirectionality, higher dependence $I_D(V_{DS})$ in saturation and lower I_{ON} than CMOS [34], resulting in a high degree of difficulty in maintaining a balance between stable *read* and *write* operations [26, 30, 31], and achieving sufficiently low access times. Therefore, TFET SRAM designs have to be investigated in-depth in order to optimize area, stability, and performance.

This chapter analyzes the applicability of TFETs in LSTP applications for longer battery life and/or energy harvesting. Section 3.2 presents the state of the art of TFET SRAM designs covering performance characteristics followed by an overview of different cell topologies. Architecture issues at the memory-array level in known TFET memory designs are analyzed and proposed solutions are reviewed.

Section 3.3 describes a Dual-Port SRAM (DPSRAM) memory architecture using an 8T-TFET Dual-Port SRAM bitcell, which overcomes array-level robustness issues (*Half-Selection (HS)* and *Write-Disturb (WD)* problems). A detailed analysis of the performance in *read* and *write* is presented including stability and response time.

Section 3.4 introduces a 3T-TFET SRAM bitcell and a TFET-CMOS hybrid SRAM memory architecture based on this cell. The detailed operation and performance are described and the physical implementation is outlined. A hybrid 128*64 bit array is analyzed and shown to have superior performance compared to a CMOS memory of the same size.

© Springer Nature Switzerland AG 2021
N. Gupta et al., *TFET Integrated Circuits*,
https://doi.org/10.1007/978-3-030-55119-3_3

3.2 TFET SRAMs - State of the Art

3.2.1 6T-TFET SRAM Design

The operation of the typical CMOS 6T-SRAM in active modes is depicted in Fig. 3.1a, b. In CMOS the *read* and *write* mechanisms can be optimized separately by tuning the corresponding transistor ratios, [35], due to the bidirectional current flow in the access transistors, see Fig. 3.1a, b. In a TFET SRAM the transfer transistors are unidirectional.

The 6T-TFET SRAM operation with outward transfer n-type transistors, as presented in [31], is depicted in Fig. 3.1c, d. Arrows in the TFET symbol show the direction of the forward current flow and hence point at the position of the source, see Fig. 2.16, Chap. 2. In *read* both bitlines are precharged at GND and the current flowing from the cell pulls up the bitline on the side storing a "1," thus creating a voltage difference ΔV between the bitlines. During *write*, the bitline BLR on the side of the node V2 being set to "1" is pulled up to VDD. In this situation, while node V1 is pulled down by bitline BLL to GND, the reverse-biased TR2 current I_{leak} which is high enough, for $V_{DS} = -VDD$, to contribute in completing the *write* operation by flipping the cell through positive feedback. Contrary to the typical CMOS SRAM, in a TFET memory the current flows through the same transistor pair

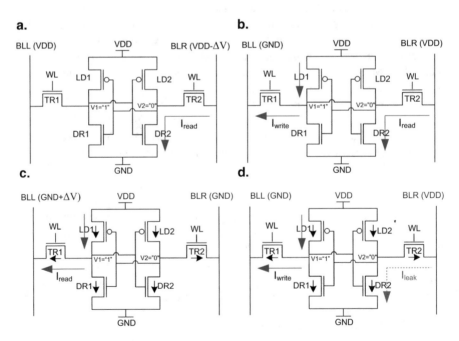

Fig. 3.1 Typical bulk CMOS 6T-SRAM operation during *read*, (**a**) and *write* (**b**); the TFET 6T-SRAM operation during *read* (**c**) and *write* (**d**) [©2009 IEEE]

(transfer and load transistors) in both operation modes. This means that the same transistor pair should allow at the same time to pull down the internal storage node to GND during *write*, and disallow this node to discharge during *read*. As a result, in the 6T-TFET SRAM design, contrary to the 6T-CMOS SRAM, it is impossible to optimize *read* and *write* stabilities separately. Therefore, a compromise between the two has to be found.

3.2.1.1   Stability

The metric used typically to characterize SRAM stability is the Static Noise Margin (SNM). Similar to Kim's data [31], the obtained *Read* SNM (*RSNM*) and *Write* SNM (*WSNM*) have values around 30 mV with optimum sizing of the cell. Such low values of noise margins are unacceptable and for this reason Kim suggested to resize the cell with a low sizing ratio (PU), pull-up (LD1/LD2) to access transistor (TR1/TR2), in order to optimize for *write* stability and use an extra transistor. The comparison of *RSNM* and *WSNM* of the bitcell as a function of the PU with the driver-NTFET size kept constant at $VDD = 1$ V is also presented. We show in the following section that unless certain architecture-level techniques are applied, this solution has several issues that are difficult to overcome if the analysis framework is enlarged and the TFET SRAM cell is placed in a memory array.

3.2.1.2   Array-Based 6T TFET SRAM Analysis

3.2.1.2.1   Introduction

Figure 3.2 depicts a nine-cell fragment of a TFET SRAM array. In the middle of this fragment is the cell accessed for *write*, ACC. The corner cells RET are in *retention* with both bitlines and the wordline set to "0." The *HS* cells, *Half-Selected* cells are connected to the same row as the written cell and are in *read* mode. The *WD* cells connected to the same column as the written cell, are under *Write-Disturb* with the wordline set to GND, one of the bitlines set to "1" and the other bitline set to "0." Left and right columns indexed "0" and "2," respectively, have bitlines initially precharged to GND and floated with BL[0] and BL[2] being slowly pulled up by *HS* cells due to "1" being stored on the left node of these cells. BL[1] and BLB[1] are connected to GND and VDD, respectively, to ensure a correct write of the "ACC" cell.

3.2.1.2.2   Half-Selection (*HS*) Problem

The problem of half-selection is a known issue for CMOS SRAM design. It consists in the cells on the same wordline as the one accessed for *write* being in the *read* condition with both bitlines precharged at VDD or GND and floated. This sets

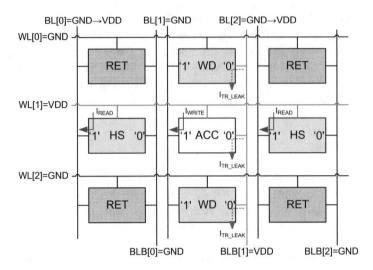

Fig. 3.2 3 × 3 bitcell array showing *HS* and *WD* problem

a number of constraints on the SRAM design. First, when sizing the cell and evaluating the stability, one has to maintain sufficient *read*-stability of the 6T cell core structure. In the design of TFET SRAMs *HS* can be an issue due to the initially low *RSNM* and *WSNM* values.

In literature, various circuit techniques are proposed for addressing these issues [26, 30–32]. However, these techniques require additional silicon area and increase the design and operation complexity. Furthermore, application of these techniques in TFET SRAMs is particularly challenging due to the device poor dynamic performance further contributing to the already low speed of TFET memories.

3.2.1.2.3   Write-Disturb (*WD*) Problem

The *Write-Disturb* problem is caused by leaky cells in the bitcell array during *write* operation. As shown in Fig. 3.3 for a 3 × 3 TFET bitcell array, the top- and bottom-row bitcells can be highly leaky depending on the bitcell data and bitline voltages while writing the center row. The leakage is due to high reverse-biased V_{DS} on the access transistors on the bitlines, which are at VDD during *write* operation. In our simulations a 32 nA bitcell leakage is measured at 1 V supply voltage. In the worst-case *WD* in a 1 Kb bitcell array with 32-bit words, where all the cells are leaky except the word being written, a 30 μA ((1024 − 32) ∗ 32 nA) leakage current is consumed.

In order to reduce the leakage caused by *WD*, supply voltages are limited to ultra-low values of less than 0.3 V. However, at such low voltages lower performance (*read/write*) and stability of bitcells become an issue. The following section presents a summary of circuits proposed to address the above issues of TFET SRAMs.

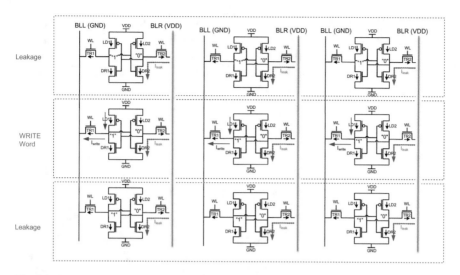

Fig. 3.3 3 × 3 TFET bitcell array showing *WD* problem

3.2.2 Other TFET SRAM Topologies

Saripalli et al. [26] published an analysis of the 6T-TFET SRAM cell showing that a 6T-TFET SRAM either with inward or outward access transistor cannot perform *read* and *write* simultaneously. They propose a number of 8T and 10T bitcells shown in Fig. 3.4 for hetero-junction TFET SRAMs simulated at $V_{DD} = 0.3$ V. All the bitcells presented are limited in supply voltage due to increasing leakage in access transistors with increasing reverse-bias V_{DS} voltage; as presented by the authors these cells can function only up to 0.3 V supply resulting in limited performance.

Yang et al. [34] focused on the analysis of the 6T cell with various kinds of transfer transistors (inward or outward NTFET or PTFET), and on the analysis of the efficacy of assist techniques for $V_{DD} = 0.8$ V. However, these cells require a virtual ground technique to perform a *write* and have a *read*-SNM close to zero resulting in unstable operation.

Kim et al. [31] presented a novel 7T architecture shown in Fig. 3.5, simulated using hetero-junction TFETs at $V_{DD} = 0.5$ V where the extra transistor serves as the *read* port, separating the *read* and *write* mechanisms, similar to the CMOS 8T SRAM cell [36]. The proposed circuit supply is again limited due to high parasitic leakage through access transistors (AXL and AXR), which would occur for a higher supply voltage due to the reverse-biased V_{DS} during *write* for *WD* cells. This may also corrupt the data in *WD* cells during the *write* operation.

Yet another 6T cell structure operating at $V_{DD} = 0.3$ V was simulated by Singh et al. [30], where the 6T cell core itself is modified to account for the particular operation mode of TFETs. The bitcell, shown in Fig. 3.6, has both access transistors connected to the same storage node. During *write* the virtual ground $W R_A$ is pulled

Fig. 3.4 8T-TFET bitcells
[©2011 IEEE]

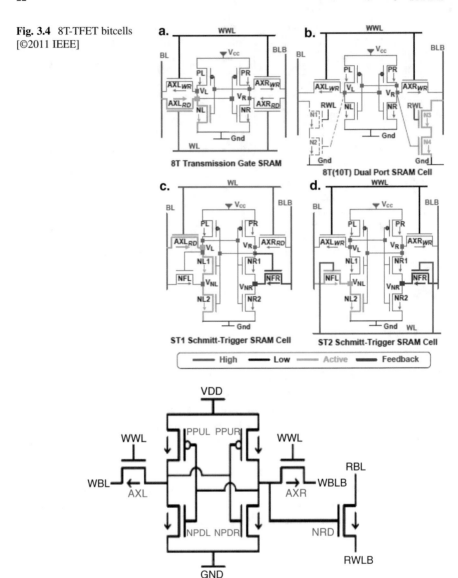

Fig. 3.5 7T-TFET bitcell [©2011 IEEE]

high and a "1" and "0" are written using M5 and M6, respectively. Since both *writes* are done using NTFETs, this results in an asymmetric *write* performance and stability. In addition, the issue of *WD* cells is still valid creating problems using this cell in a memory array. Therefore, the operating voltage is limited for this cell as well.

Fig. 3.6 Modified 6T-TFET
bitcell [©2010 IEEE]

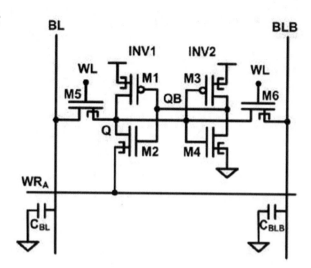

Fig. 3.7 5T-TFET-CMOS
hybrid bitcell

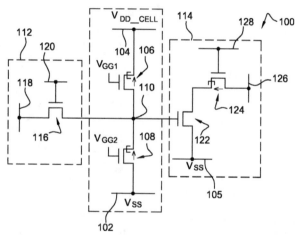

In [37] a 5T-TFET SRAM bitcell is presented using the *Negative Differential Resistance (NDR)* property of TFETs, see Fig. 2.5, Chap. 2. Figure 3.7 shows the low-voltage SRAM bitcell implementation using two NTFET devices, which are connected in series with reverse-biased V_{DS}. Storage node 110 has two stable states based on which of the two device currents is much larger than the other one. While storing "0," device 108 is in *NDR* state having reverse-biased $V_{DS} \approx 0$ V with 106 in OFF state. Similarly, for storing "1," device 106 is in *NDR* state having reverse-biased $V_{DS} \approx 0$ V with 108 in OFF state. The *read* port is implemented using one MOSFET and one NTFET to have high *read* current without impacting the bitcell stability by isolating the storage node from the *read* current. In order to write "0" and "1" using a single device the *write* port is implemented by MOSFET 116; it would not be possible to use a TFET due to its unidirectional property. The bitcell

Fig. 3.8 4T-*NDR* TFET
bitcell

Fig. 3.8 4T-*NDR* TFET
bitcell

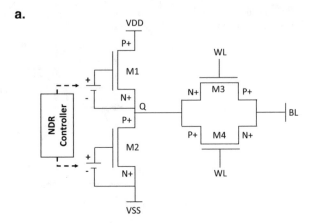

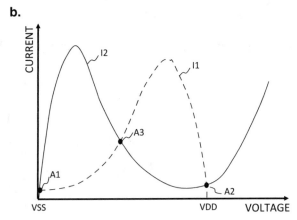

stability constraint consists in ensuring that the write port MOSFET (116) leakage is lower than the TFET (106, 108) *NDR* current. Depending on the technology this may require using a higher threshold voltage device, longer channel length or, if available, use reverse body biasing. While each of these solutions are easy to implement in the design, they could lead to increased area and physical design complexity.

In [38] another TFET SRAM bitcell is presented using the *NDR* property of TFETs. Figure 3.8a shows the low-voltage SRAM bitcell implementation using two NTFET devices, M1 and M2, which are connected in series with reverse-biased V_{DS}. The *read* port is implemented using two TFETs due to the unidirectional property. The *NDR* device currents $I1$ and $I2$ of M1 and M2, respectively, are plotted vs. Node Q voltage in Fig. 3.8b. Points A1 and A2 are two stable states where one device current sensitivity to voltage change is much larger than that of the other keeping the voltages at node Q stable. This 4T-TFET bitcell has *read* stability issues due to the low magnitude of *NDR* currents in TFETs (in the pA range). Therefore, in order to not disturb the data in the cell a *read* should be done with low *read* currents in *pA* range resulting in very slow operation.

In summary, all the above-mentioned published TFET SRAMs revealed difficulties in obtaining sufficient stability in *read* and *write* operations. As the stability in both operation modes is inherently low due to the electrical performance of TFETs and the low supply voltage, it is difficult for circuit designers to find the best balance between *read* and *write*. Moreover, due to the unidirectional TFET behavior researchers were forced to target very low-V_{DD} operation resulting in even larger difficulty in achieving sufficient stability margins in active mode. New architectural solutions were developed to improve the single-cell stability but the proposed cells in [30, 31] suffer from *HS* and a special case of *WD* when organized as an array of memory cells. Similar to TFET SRAMs, overcoming *HS* in CMOS SRAMs is particularly important for 8T CMOS cells with full separation of *read* and *write* mechanisms.

Two solutions were proposed in literature for the 8T CMOS SRAM: (1) all bits on the selected wordline must be spatially adjacent (no bit interleaving) with the wordline length covering only these bits [39] and (2) the implementation of a column-based *write-back* mechanism [40, 41]. These solutions could be applied to TFET SRAMs also, but they will result in an area penalty and in additional design and operation complexity. Additional operation and design complexity should be avoided for devices with rather poor dynamic performance such as TFETs. In addition, the design with no bit interleaving does not allow the application of conventional Error Correction Codes (ECC), which is essential to handle soft errors [42]. It should be noted that the Soft-Error Rate (SER) increases significantly under scaled-down V_{DD} operation [43], which is commonly used in TFET-based designs to avoid high current due to the turn-on of the p-i-n diode with high reverse-biased V_{DS}. As shown in [44, 45], TFETs are better in terms of soft-error rate; however, for circuits with both CMOS and TFETs the soft-error rate increases with scaling down the supply voltage. This necessitates the application of an efficient ECC scheme resulting in further increase in design overhead, complexity, and area. In [39], the authors propose a long column for write (512 cells) with the advantage of a short wordline in designs with no bit interleaving. This choice can cause a number of issues in TFET designs due to the *WD* problem. An 8T-TFET SRAM with dual wordlines and consuming two clock cycles for writing a single word, which is *HS* and *WD*-free at the cost of higher area (similar to a dual-port memory) is presented in [32]. As shown in Fig. 3.9, the bitcell is designed such that none of the TFETs experiences a high reverse-biased V_{DS}. The advantage of this cell is that it can provide low-leakage operation at high supply voltages. This circuit is free from *HS* and *WD* issues at the cost of dual wordlines and three supply voltage levels. In addition, both "0"s and "1"s cannot be written at the same time in the memory array and need two *write* cycles for storing one word.

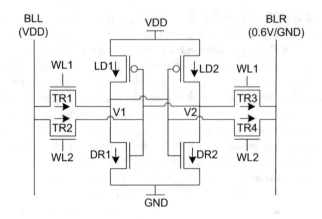

Fig. 3.9 8T-TFET dual-wordline bitcell [©2012 IEEE]

3.3   Dual-Port SRAM Architecture

DPSRAMs provide simultaneous access to a single static SRAM memory from
two buses with full synchronous operation on both ports. In embedded systems
the use of dual-port memories is increasing in order to optimize area, power, and
throughput requirements. As an example, dual-port SRAMs are used as buffers
in DSP processors [46]. Multi-port SRAMs are used as scratchpads to provide
independent memory access to each core while using shared memory resources
[47, 48]. With increasing demand for performance and feature size shrinking, power
consumption in SRAMs is rising. This problem is critical in DPSRAMs due to the
added complexity in the cell. In a standard 8T-CMOS dual-port SRAM [46] cells
need to be larger to get the same stability as single-port SRAM, because in a worst-
case situation two *read* operations can occur on the same row. This event results in
a $2\times$ current drive increase of the pass transistor reducing the *RSNM* and requiring
the increase of the SRAM cell size to improve stability. In [47], *write* assist is
used to reduce the minimum supply voltage required for correct operation. Another
common method used to optimize power for different performance requirements in
DPSRAMs is Dynamic Voltage Frequency Scaling (DVFS) [49]. However, DVFS
only partially addresses the leakage power consumption in SRAMs as leakage
power is the dominant part in the overall power consumption. Thus, optimization
at cell and/or architecture level is required for achieving low leakage and power
efficiency. A 15 μA/Mb leakage is reported at 0.7 V supply in 45 nm CMOS in [49],
and in [8], the authors propose a 6T-SRAM cell of $2.159\,\mu m^2$ using long-channel
and thick-oxide devices; with the application of reverse back-bias (RBB) during
standby authors report over $10^3\times$ leakage reduction (27 fA/bit) at room temperature
as compared to a standard SRAM design.

The following section presents a TFET-based 8T-Dual-Port SRAM cell [50] and
TFET/CMOS hybrid memory architecture with ultra-low leakage for scratchpad

Fig. 3.10 Block diagram of
proposed dual-port
scratchpad [©2015 IEEE]

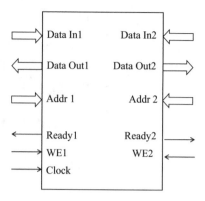

application. The design of the pseudo dual-port scratchpad is optimized for both speed and power. SRAM arrays are major contributors to overall leakage power consumption and, contrary to peripheral circuits, can never be power gated; therefore, in order to limit leakage the memory cells are designed using TFETs. However, TFETs have low current drive and high capacitance as compared to CMOS and therefore, they take more time to charge-discharge the nodes and consume more switching power. In order to optimize speed and power, logic and drivers for high-capacitance nodes with high switching activity are designed with CMOS. Thus, all the periphery circuits such as WL drivers and sense amplifiers, which have greater switching activity are designed with CMOS at the cost of increased leakage. Since CMOS is having more current drive as compared to TFETs, CMOS drivers are smaller in size for a given speed specification.

Figure 3.10 shows the block diagram of the proposed scratchpad using DPSRAM with both ports having *read* and *write* feature. In the proposed design, two simultaneous *read* or a single *write* on either port is supported. This is done to optimize the cell because the number of *reads* are significantly larger than the number of *writes* in a DPSRAM used as scratchpad for embedded processors; for DSP applications such as video decoding, once the frame is decoded and saved in the SRAM buffer, the same frame buffer may be read several times for motion estimation, predictions and to compute other frames. For this purpose, the proposed scratchpad design has a "Ready" signal per port to notify that memory is ready to accept a new *read* or *write* operation. Since a single *write* is allowed, if a *write* is ongoing on port-1, the "ready" signal on port-2 will be "0" to notify the processor that the memory is busy.

3.3.1 Dual-Port TFET SRAM Cell

The proposed TFET DPSRAM cell [50] is shown in Fig. 3.11; it has two wordlines (WL1 and WL2), two bitlines (BLR1 and BLR2) used for *read* and *write*, and two bitlines (BLW1 and BLW2) used for *write*. A single-ended read scheme is applied

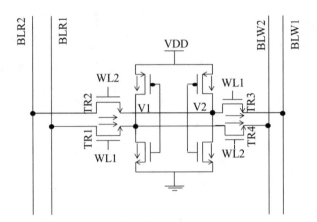

Fig. 3.11   8T-TFET dual-port SRAM cell [©2015 IEEE]

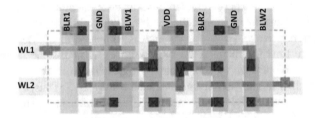

Fig. 3.12   8T dual-port cell layout with two bitline connections. Top connections made with Metal-2 for BLR1&2, BLW1&2, VDD, GND (vertical) and Metal-3 for WL1 and WL2 (horizontal) [©2015 IEEE]

on each BLR1 and BLR2 separately to allow two simultaneous *read* operations. The unidirectional behavior of the device is satisfied by using only BLR1 or BLR2 for *read*. The voltage range is restricted on bitlines during active operation modes.

In *retention* the BLRs are set to VDD and the BLWs are kept at 0.6 V. Under such conditions if node V2 stores a "0," TR3 operates in reverse at $V_{DS} = -0.6$ V and the TFET current remains in the fA range, therefore not causing a significant leakage. In this case, the other transfer transistor connected to the BLW2 (TR4) has its drain at the node storing "1" and hence operates at a positive V_{DS} where the current for $V_{GS} = 0$ V is always very low. This approach allows maintaining a negligible leakage for *WD* cells. For the other case when node V2 stores a "1," the total leakage will be the same with TR3 operating in forward bias and TR4 in reverse bias. Due to single-ended *read* operation bitline-multiplexing logic is half of the required logic for differential sense amplifiers. This optimization is important because normally the bitline-multiplexing logic is double for DPSRAM as compared to Single-Port SRAM (SPSRAM).

Figure 3.12 shows the physical implementation of the cell. It is similar to the standard 8T Dual-Port (DP) cell in [48]. It should be noted that in dual-port cells

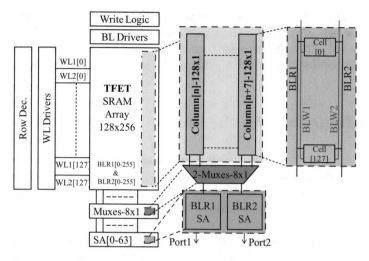

Fig. 3.13 32 Kb Mux-8 dual-port memory architecture [©2015 IEEE]

each access transistor is connected to a separate bitline and therefore the layout has four physical bitline connections. The cell area is evaluated at $0.338\ \mu m^2$ in a 28 nm FDSOI process, with a 29% area overhead as compared to an industrial cell layout of a 6T-SPSRAM cell as reference; the same transistor sizing is used for both but the latter does not have the two additional access transistors.

3.3.2 Micro-Architecture Design

The architecture of the 32 Kb TFET dual-port scratchpad memory [50] is depicted in Fig. 3.13. The memory is organized in a single array of 128 rows and 256 columns with a 32-bit word and bit interleaving. Therefore, each set of eight columns has two 8×1 read MUX's and two sense amplifiers to enable dual-port *read*. The memory array is built using 8T-TFET DPSRAM cells, Fig. 3.11, and the peripheral circuits are designed with 28 nm FDSOI CMOS devices.

3.3.3 Read Operation

The *read* operation is similar to that of a typical 6T-SRAM, with the difference that it is single ended. Sensing depends on the selected *read* bitline, BLR1 or BLR2, being discharged through the cell when "0" is stored on the corresponding node, V1 or V2, of the memory cell, or remaining at VDD when the value of the read node is "1." The minimum (critical) *WordLine Pulse* width, WLPcrit, required to pull down the BLR voltage by 150 mV is 377 ps at 1 V supply. This results in less than 1*ns* *read* cycle time. Single-ended sensing with long bitlines or open bitline architecture with split bitlines can be used for sensing depending on the memory architecture choice. The next section describes both of these single-ended sensing schemes.

3.3.4 Single-Ended Sensing Methods

In order to optimize the TFET cells single-ended sensing is used in our designs. Single-ended sensing has already been reported in literature [51–53]. Single-ended sensing can be done using a differential sense amplifier with a reference voltage source or with inverter-based sensing. In the case of inverter-based sensing, the bitline needs to be discharged fully resulting in slow operation. In [52], an inverter-based sense keeper is implemented in the *read* circuit to improve the *read* speed. After analyzing various single-ended sensing methods we designed single-ended sensing using an imbalanced differential sense amplifier based on charge injection. Our design uses charge injection to create the imbalance for removing the reference voltage source and symmetric sizing of devices for minimizing variations. The proposed sensing technique used in this SRAM is presented in Chap. 7, Sect. 7.2.

3.3.5 Write Operation

In order to perform the *write* operation in the proposed 8T cell, see Fig. 3.11, both WL1 and WL2 are set high. Depending on the value to be written, either BLW1 or BLW2 is pulled down to GND and the other remains at the *retention* voltage, i.e., 0.6 V. Also, the BLR corresponding to the same wordline as the BLW, which is pulled down remains high, and the other BLR is set to 0.4 V. For instance, in order to write a "1" at node V1 in the cell, BLW1 is 0 V, BLW2 is 0.6 V, BLR1 is VDD, and BLR2 is 0.4 V. Figure 3.14 shows the *write* waveform in the cell. BLR and BLW for *HS* cells are at the *retention* voltage. This configuration of voltages on bitlines ensures that the TFET pass transistors have minimum leakage for *HS* and *WD* cells.

Contrary to the typical 6T-TFET single-port SRAM cell, in our 8T cell no transfer transistor is reverse biased close to the turn-on of the p-i-n diode in the memory active mode resulting in no impact of *WD* cells on performance, stability, and leakage. The analysis of different *write*-assist techniques for TFET cells shows that WL boosting reduces noise margins in *HS* cells, which are in *read* mode. *Write* time increases drastically with supply under-drive (Vddud) for TFET SRAM cells because of the steep subthreshold slope of TFETs and non-saturating $I_D(V_{DS})$ dependence. The *write* margin also limits the minimum operating voltage. The TFET pull-up in the SRAM cell is very weak at lower voltages and takes much longer to charge the memory cell node resulting in a very long critical wordline pulse width (WLPcrit) requirement for a successful *write*. A comparison of WLPcrit

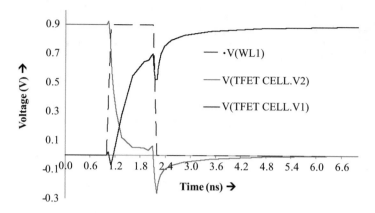

Fig. 3.14 *Write* waveform with 0.9 V Supply [©2015 IEEE]

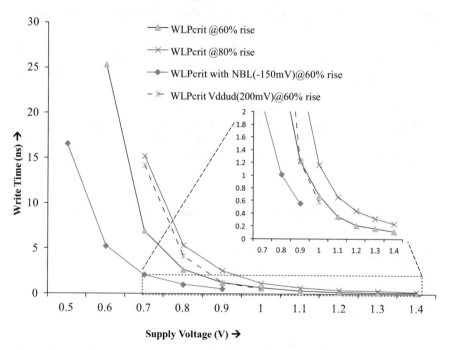

Fig. 3.15 Comparison of *write*-assist techniques [©2015 IEEE]

for different *write*-assist techniques was performed; the resulting WLPcrit without *write* assist and with 150 mV negative bitline (NBL) and Vddud *write* assist is shown in Fig. 3.15. WLPcrit with *write* assist is shown for the voltage rising to 60% of VDD, which for the proposed 8T-TFET DPSRAM cell is sufficient to execute a *read* successfully on the same cell in the next cycle. Figure 3.15 also shows WLPcrit for *write* without assist, the difference between WLPcrit for 60% and 80%

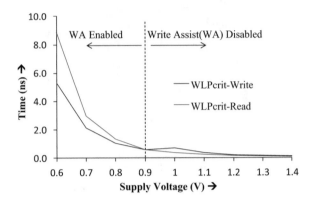

Fig. 3.16 Minimum wordline pulse width requirement vs. memory supply [©2015 IEEE]

VDD rise of internal node voltage is due to the non-saturating $I_D(V_{DS})$ of TFETs. The negative bitline (NBL) seems to be best suited for TFET SRAM cells in order to reduce VDD and maintain speed. TFETs used for implementation have V_{OFF} voltage of 150 mV as described in Chap. 2. Because of very-low leakage current in TFETs for $V_{GS} < V_{OFF}$, the power consumption penalty and stability issue of *WD* cells is negligible with negative bitline *write* assist down to -150 mV. With 150 mV negative bitline, 79.2% to 54.7% performance improvement was achieved for a supply voltage range from 0.6 V to 0.9 V, respectively. Without *write* assist WLPcrit for *write* is more than 40 ns for a supply below 0.6 V while with *write* assist enabled WLPcrit is less than 17 ns at 0.5 V supply voltage. Above 0.9 V supply voltage WLPcrit is less than 1 ns with *write* assist disabled (see Fig. 3.15).

3.3.6 8T-TFET SRAM Stability and Performance

The performance and stability of the 8T dual-port TFET SRAM was analyzed for a range of supply voltages for applying DVFS. Figure 3.16 shows the WLPcrit for *write* and *read* operation of the designed 32 Kb memory; in order to keep the delay at reasonable values *Write-Assist (WA)* techniques are applied if VDD is lowered below 0.9 V. The achieved WLPcrit for *read* and *write* at 1 V supply is 681 ps and 377 ps, respectively.

Figure 3.17 shows the *RSNM* and *WSNM* for different supply voltages. The achieved noise-margin values at 1 V supply are 114 mV for *RSNM* and 185 mV for *WSNM*. An improvement of approximately 5× and 8× in *RSNM* and *WSNM*, respectively, is observed in comparison to computed margins for a standard 6T TFET SRAM cell with outward access transistors in the balanced case [31]. The achieved *RSNM* and *WSNM* are sufficient for supply voltages of 1 V and above. At lower voltages noise margins can be increased by applying assist techniques for *read* and *write* resulting in higher *RSNM* and *WSNM*, which enable the use of long wordlines with bit interleaving without any *HS* problem. In a TFET the leakage

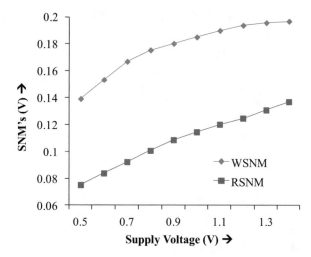

Fig. 3.17 Static noise margins (*Read/Write*) vs. memory supply [©2015 IEEE]

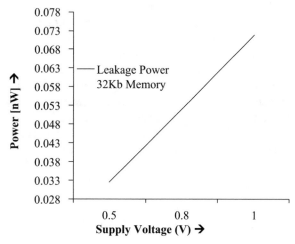

Fig. 3.18 Leakage power vs. memory supply [©2015 IEEE]

current is 10^5 times smaller than that of a CMOS and is independent of supply voltage, thus the leakage power of the designed TFET SRAM cell varies linearly with supply voltage. We achieved a leakage of less than 5 fA/bit for a supply range of 0.6 V to 1.4 V. Figure 3.18 shows the leakage power consumption extracted from simulations of the designed memory cell array. The results presented in [50] show that TFET circuits can be competitive with LP CMOS also in terms of speed, as long as larger than nominal supply voltages are used. In the described 8T-TFET dual-port cell the leakage is carefully controlled by avoiding excessive reverse biasing of any of the devices resulting in a very low increase of cell leakage and higher operating frequency with increased supply voltage.

3.4   3T-TFET Bitcell-Based TFET-CMOS Hybrid Memory

In Sect. 3.3, a TFET/CMOS hybrid Dual-Port SRAM (DPSRAM) based scratchpad memory was presented with ultra-low leakage current (<5 fA/bit) and 29% increased area in comparison to an industrial 6T-CMOS SRAM. However, dynamic power consumption for these memories is more than that of a 6T-CMOS because of increased load on high-capacitance nodes, such as wordlines. An alternative is offered in [37, 38], where TFET latches and memory cells using the *NDR* property of TFETs in reverse bias are presented. The *NDR* property of TFETs [14] is very promising for designing a compact latch; however, the architecture proposed in [38] suffers from stability and performance issues. In order to maintain data during *read*, the *read* current should be less than the hump current (in pA range) provided by *NDR*. This constraint leads to an extremely slow *read* with the risk of data corruption while executing the operation; additionally, the TFET transmission gate for data access limits the maximum operating voltage.

This section provides insight into an ultra-compact SRAM design using Si-TFETs [14] compatible with CMOS for ultra-low power applications with ultra-low leakage for long battery life time and good performance. We analyzed the architecture-level issues in TFET SRAM design and demonstrate a novel 3T-TFET SRAM cell designed using the *NDR* property of TFETs in reverse bias. The proposed design supports aggressive voltage scaling without impacting data stability of the cell and allows the application of performance-boosting techniques without impacting cell leakage. The new cell maintains reasonable stability in all operation modes without using any assist technique. Based on this 3T-TFET bitcell a TFET/CMOS hybrid memory architecture is proposed using CMOS peripheral circuits.

3.4.1   Proposed 3T-TFET SRAM Cell

The key concept of the 3T-TFET SRAM bitcell [54] based on a static latch using *NDR* with two TFET devices, one NTFET (M1) and one PTFET (M0) is shown in Fig. 3.19. The two supplies VD and VS assume different values depending on the mode of operation of the cell.

During *retention* the voltages on VD and VS insure that devices M0 and M1 are reverse biased with $0 < V_{VD} - V_{VS} \leq 0.6$ V and BiasM0/BiasM1 are chosen such that both devices get sufficient gate drive for large enough hump currents, see Chap. 2, Fig. 2.5. The I_D vs. V_{Qint} characteristics for the two TFET devices M0 and M1 connected in series with reverse biased V_{DS} applied are shown in Fig. 3.20. This transistor configuration biased as above results in a latch behavior with the condition that the total cell supply be limited to the critical value where the TFET current becomes independent of gate voltage; for our devices this point is at 0.6 V.

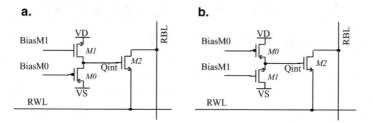

Fig. 3.19 Proposed 3T-SRAM cell architectures [©2016 IEEE]

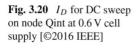

Fig. 3.20 I_D for DC sweep on node Qint at 0.6 V cell supply [©2016 IEEE]

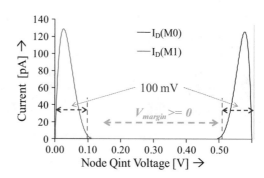

The *read* operation is performed using the RBL and RWL lines with RBL precharged and RWL active low. The *write* operation is performed using a combination of voltages on VD, VS, and BiasM0/BiasM1, which enables storing a "0" or a "1" in the cell on node Qint. The various operating modes are described in the following subsections.

Two versions of the proposed cell are shown in Fig. 3.19a, b. The circuit shown in Fig. 3.19a provides high *write* speed and low capacitance on the supply nodes VD and VS because of a low C_{GS} in TFETs and constant V_{GS} during *write* for M0 and M1. In the circuit shown in Fig. 3.19b, having the two TFET sources connected together at Qint, V_{GS} for M0 and M1 is continuously changing with V_{Qint} during the *write* operation resulting in a performance penalty. However, this circuit is better in terms of stability due to higher hump current and higher V_{GS} for M0 and M1 during retention mode.

3.4.1.1   Retention Mode and Stability

The information in the cell is stored on node Qint during *write* by forward biasing one of the transistors and turning the other one off. Figure 3.20 shows I_D of M0 and M1 as a function of V_{Qint}; BiasM0/BiasM1 for the circuits shown in Fig. 3.19 in retention have values such that the TFETs are turned on in reverse-V_{DS} mode. The state of the cell "0" and "1" set during a *write* is represented by one of the two humps. As both TFETs operate in the *NDR* region, M0 preserves the "0" (V_{VS})

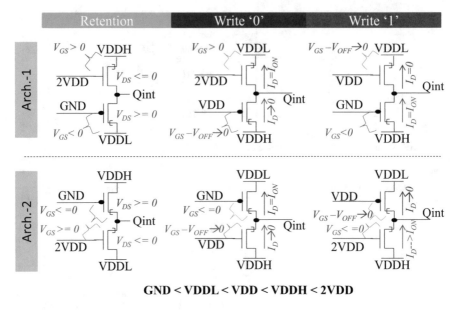

Fig. 3.21 TFET terminal voltages for different cell states [©2016 IEEE]

and M1 preserves the "1" (V_{VD}) controlled by BiasM0 and BiasM1, respectively. For a cell supply VDD = 0.6 V node Qint is kept discharged by M0 for $0 < V_{Qint} < 100$ mV ('0'), and Qint is kept charged by M1 for 0.5 V $< V_{Qint} < 0.6$ V ("1"). Terminal voltages of the two TFETs during *retention* and *write* are shown in Fig. 3.21.

V_{margin} measures the voltage difference between the two stable states of the cell and the voltage range, for which the cell is metastable. The reverse-current peak value varies with the applied gate voltage but the width of the hump remains fairly independent of gate voltage. The stability constraints for the proposed cell are significantly different from a conventional 6T-SRAM cell because the data storage node Qint is isolated in all operating conditions, except in *write*. Therefore, stability during *read/write* operation is similar to the static noise margin (SNM) of the cell resulting in a weak dependence of the SNM on the cell's supply voltage. The cell SNM is 100 mV, equal to the width of the current hump for $V_{margin} \geq 0$.

3.4.1.2 Write Operation

During *retention* M0 and M1, see Fig. 3.19, are reverse biased; in order to *write* into the cell shown in Fig. 3.19a (Arch-1), the supplies connected at VD and VS are swapped to make $V_{VS} > V_{VD}$. Terminal voltages of the two TFETs during *retention* and *write* are shown in Fig. 3.21.

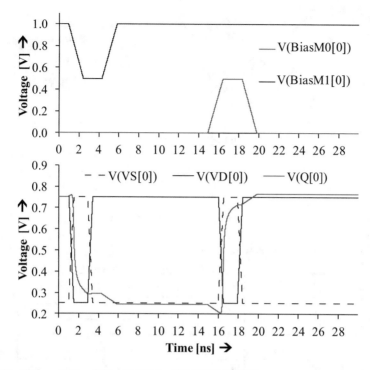

Fig. 3.22 *Write* operation (Write-"0," Write-"1") [©2016 IEEE]

For writing a "0" in the cell, BiasM0 is pulled up from its retention value to VDD to reduce the gate drive of M0 and BiasM1 remains at the *retention* voltage; thus, M1 discharges the node Qint to the voltage on node VD. Similarly, for writing a "1," BiasM1 is pulled down to reduce the gate drive of M1 and BiasM0 remains at the *retention* voltage; therefore, M0 charges the node Qint to the voltage on node VS. The simulated waveforms for writing "0" and "1" in the cell are shown in Fig. 3.22.

3.4.1.3   Read Operation

Read operation is performed based on a single-ended read scheme using the RWL and RBL lines. RWL selects the row to be *read* by taking it low and RBL either discharges or remains at a precharged voltage depending whether the data in the bitcell-node (Qint) is "1" or "0," respectively. RBL can be allowed to discharge fully or a single-ended sense amplifier can be used; full discharge is preferable for low-voltage operation. An inverter-based sensing circuit can be used for this purpose, which makes it easier to match the reading circuit layout pitch with that of the bitcell column pitch. The *read* waveform is shown in Fig. 3.23 for reading "0" and "1."

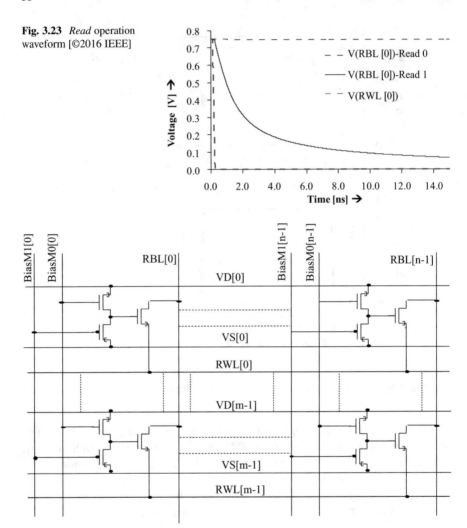

Fig. 3.23 *Read* operation waveform [©2016 IEEE]

Fig. 3.24 Bitcell and corresponding signal organization in mxn-SRAM cell array [©2016 IEEE]

3.4.2 Memory Architecture

Figure 3.24 shows the memory cell array organization including the routing of signals. Data words are stored horizontally, the VD and VS lines are routed horizontally to align with the data word and BiasM0 and BiasM1 are routed vertically. In this architecture selection of the row to be written to is done by VD and VS by swapping their values from the ones in retention, see Fig. 3.21; the value to be written in each cell is decided by BiasM0 and BiasM1. Selection of the row to be *read* is done by RWL and data is *read* using RBL's.

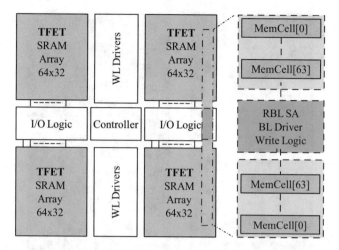

Fig. 3.25   Proposed TFET/CMOS hybrid memory architecture [©2016 IEEE]

The proposed memory architecture is shown in Fig. 3.25. In order to optimize the cell array leakage current the bitcell array is designed only with TFETs while the periphery uses CMOS to optimize the area for the same speed of operation due to its higher drive strength when compared to that of TFETs. We have used the single-ended skewed inverter-based sense amplifier for reading, described in Chap. 7, Sect. 7.4.1, to limit the bitline discharge, to reduce power consumption, and to allow for a larger column size.

3.4.3   Memory Layout

The layout for the dual-cell block and the 64 × 32 cell array is shown in Fig. 3.26. The cell size is $0.1266 \, \mu m^2$/bit using 28 nm FDSOI process logic design rules; this area is similar to that of an industrial high-density 6T-CMOS cell, which is implemented using compact design rules instead of logic design rules. In this layout the TFET M2 (200 nm) on the *read* port is twice the size of M0 and M1 (100 nm). This improves the *read* speed of the design. In order to achieve an optimized rectangular layout of the memory cell, two bitcells are layed out together to have six transistors within a dual-bitcell layout structure, see Fig. 3.26. Cell boundaries are represented by dotted lines to show each cell separately. Due to the reduced cell width the wiring capacitances on the various horizontal lines in the cell array are reduced. The capacitance values of metal lines extracted from the layout represent only 50% of those of an identical-size memory array designed using the compact 6T-SRAM cell. This reduction combined with a low C_{GS} capacitance of TFET devices cuts the total capacitance on VD, VS, and RWL lines to less than half resulting

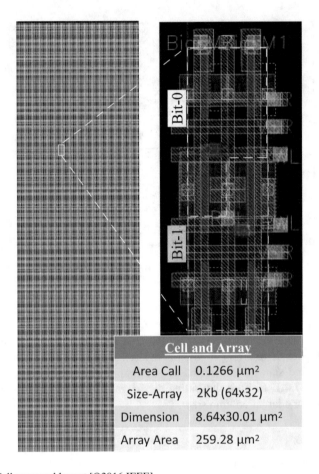

Cell and Array	
Area Call	0.1266 μm²
Size-Array	2Kb (64x32)
Dimension	8.64x30.01 μm²
Array Area	259.28 μm²

Fig. 3.26 Cell array and layout [©2016 IEEE]

in drivers for RWL, VD, and VS with less leakage for the same specification of transition time and word size.

3.4.4   Energy Efficiency

The energy consumption of the proposed design is computed with the assumptions that during *read* 50% of data are "0" and 50% are "1," and 50% of all operations are *read* and 50% are *write*. An overall comparison of performance data for the presented memory cells including energy consumption in various operation modes and bitcell area is shown in Table 3.1. E_{READ} is the energy consumed during *read* on row drivers and bitlines. Bitline discharge is limited to 200 mV for 3T-TFET and 8T-TFET SRAMs because they use single-ended sensing while it is limited

Table 3.1 Comparison - power and area [©2016 IEEE]

Cell	VDD (V)	WL_{Pulse} Read (ns)	WL_{Pulse} Write (ns)	E_{READ} (fJ/acc.)	E_{WRITE} (fJ/acc.)	E_{AVG} (fJ/acc.)	I_{LEAK} Active (pA/bit)	I_{LEAK} STBY (fA/bit)	Area-Bitcell (μm^2)
8T-TFET	1	0.26	1.10	28.5	61.0	44.8	25.5	5.00	0.336
6T-CMOS	0.75	0.27	0.16	9.10	12.9	11.0	11.8	$7.8*10^3$	0.120
3T-TFET [proposed]	0.6	0.21	0.93	1.81	4.99	3.40	5.72	0.35	0.1266
6T-CMOS [8]	1.2	7 ns (access time)[a]		N.A.	N.A.	25 μW/MHz[a]	N.A.	27.0	2.04

[a]Measurement values reported for full memory [8]

to 100 mV for 6T-SRAM with differential *read*. E_{WRITE} is the energy consumed during *write*. I_{LEAK} in active mode is the total leakage in the bitcell array and periphery with dynamic power gating (only 25% of the drivers are switched ON depending on the accessed address). I_{LEAK} in standby mode is computed with the periphery OFF and the bitcell power ON to retain the data.

The leakage power per driver of a row of the 3T TFET cells is 52% and 85% less in comparison to that of the 6T-CMOS and 8T-TFET SRAMs, respectively. Three drivers per row are required in this design resulting in 42% increased leakage per row in comparison to the 6T-CMOS high-density (HD) SRAM; however, the driver leakage per row is still 77% less in comparison to the 8T-TFET SRAM, even with two drivers per row, which have high capacitance and longer wordlines due to a larger bitcell size. The bitcell array leakage is $10^4\times$ and $77\times$ lower compared to that of the 6T-CMOS (HD) [18] and ultra-low leakage 6T-65 nm CMOS SRAM [8] cells, respectively. Bitcell leakage is $14\times$ lower in comparison to 8T-TFET SRAM cells [32]. During standby the total leakage is coming from the cell array, thus TFET memory leakage is much lower than that of CMOS memories. Overall memory leakage during active mode including bitcells and drivers for the proposed design is 52% lower than for 6T-CMOS and 77% lower than for 8T-TFET SRAMs.

Dynamic power consumption of the 3T-TFET cell design is 70% less in comparison to the 6T-CMOS SRAM and up to 90% less in comparison to the 8T-TFET SRAM due to the low capacitance on VD, VS, and RWL lines.

3.4.5 Read and Write Performance

Read and *write* minimum wordline pulse width (WLPcrit) is plotted as a function of supply voltage in Fig. 3.27. For the full range of operation cell leakage is <0.35 fA/bit because devices M0 and M1 are in reverse bias during retention. As shown in Fig. 3.27, the *read/write* speed can be increased by using assist techniques for low-voltage operation. Since the cell data storage node is isolated from RBL and the bias voltages of M0 and M1, negative wordline (NWL) can be used to increase *read* speed without impacting the cell stability. Similarly, for *write* the speed can be increased without impacting cell stability by boosting bias voltages BiasM0 and BiasM1. *Read* performance is improved by $29\times$, by using NWL of -100 mV and -150 mV for supply voltages of 0.4 V and 0.3 V, respectively, on *read* wordline RWL. A boost of 100 mV on the gate-bias voltage of M0 and M1 results in a $4.8\times$ improvement at 0.3 V cell supply for *write* operation.

The overall performance is estimated including periphery delays in the row decoder, drivers, and sensing. The proposed design supports overall *read* speed from 1.92 GHz to 3.82 MHz and *write* speed from 429 MHz to 17.3 MHz for 0.6 V to 0.3 V cell supply voltages, respectively, with corresponding BiasM0 and BiasM1 values from 1.2 V to 0.6 V. Therefore, this implementation requires a total of five voltages, which can be implemented either using five supplies or three supplies with two voltage dividers inside the memory.

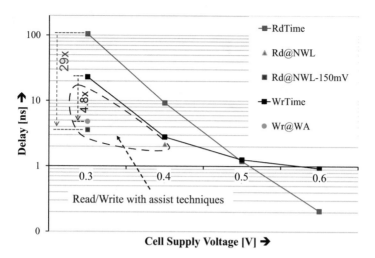

Fig. 3.27 *Read/Write* performance (WLPcrit) vs. cell supply voltage, including improvements with *read/write*-assist techniques [©2016 IEEE]

3.5 Summary

This chapter presents TFET/CMOS hybrid SRAM solutions for both typical-voltage/high-performance and ultra-compact/low-voltage/low performance applications. An 8T-TFET bitcell-based Dual-Port SRAM is proposed, which can work from 0.6 V to 1 V providing ultra-low standby power while maintaining good performance and stability. The achieved leakage is less than 5 fA/bit by controlling voltage bias to limit high reverse-bias TFET parasitic current. The WLPcrit of 377 ps is achieved for a supply voltage of 1 V and *write* speed improvements up to 79.2% are obtained with NBL *write assist* of 150 mV. The presented design uses four voltages, which can be optimized to three voltages either by limiting the highest voltage of operation to 0.9 V or, at the cost of increased leakage at 1 V operation.

For low-voltage applications, a novel integrated 3T-TFET bitcell-based TFET/C-MOS hybrid SRAM architecture was introduced in this chapter; TFETs are used for the memory array due to ultra-low leakage and CMOS for the periphery. The proposed 3T-TFET bitcell uses an *NDR*-based 2T-TFET latch for data storage and 1 TFET for the *read* port; it implements a new *write* mechanism using supply lines. The proposed design supports voltage scaling and works from 0.6 V to 0.3 V bitcell supply voltages; memory cells have an ultra-low leakage current less than 0.35 fA/bit. *Read/write* speed improvement of up to 70% and 90% can be achieved using assist techniques for 0.4 V and 0.3 V cell supply, respectively.

Chapter 4
Ultimate-D/SRAMs/CAMs

4.1 Introduction

On-chip storage demand is continuously increasing due to the higher complexity of SoCs. It is important to scale memory aggressively in order to increase on-chip memory capacity in a cost-efficient manner. In the past, technology scaling was key to augment the on-chip SRAM. However, at new technology nodes challenges such as increasing leakage and variations are limiting scaling SRAM capacity further. Therefore, researchers are exploring alternative options, such as *Embedded-DRAMs* (eDRAMs) and *Embedded-Flash* (eFlash), to improve memory density.

The use of eFlash is limited due to its reduced *write* endurance and higher operating voltage than that of standard CMOS digital logic. eDRAMs are a promising alternative, however, they have their own limitations such as refresh, and a costlier process due to the vertical storage capacitors. There have been several reports in literature focusing on eDRAMs to improve the array density and to reduce cost [55–58]. An intermediate solution for deploying eDRAM is to reduce the density in eDRAMs compared to standard DRAMs and thus enabling implementation in a standard CMOS process for digital logic [58–60]. However, technology raises challenges in scaling the standard 1T1C DRAM structure because of difficulties in scaling the capacitor as it needs a high value to limit the refresh rate and reduce the throughput penalty. The ITRS roadmap [3] shows a 20% reduction of the required DRAM cell capacitance to store 1 bit from year 2009 to 2016 whereas in the same time period transistor technology scaled by 57% from 52 nm to 22 nm. Various techniques such as negative wordline (NWL) and high-oxide-thickness capacitors are used in DRAMs to reduce leakage and thus to increase *retention* time. eDRAM capacitors in [55] are using an *effective oxide thickness (EOT)* of 0.7 nm to get 8 fF/bit capacitance with 0.1 fA/bit leakage. However, the EOT of 0.3 nm, suggested by ITRS for DRAM capacitors [3], would result in significantly increased capacitor leakage. In eDRAMs capacitor size is reduced at the cost of *retention* time in order to optimize cost of process and silicon footprint. In [58], a 14.2 fF/bit capacitance

© Springer Nature Switzerland AG 2021
N. Gupta et al., *TFET Integrated Circuits*,
https://doi.org/10.1007/978-3-030-55119-3_4

is implemented in eDRAM with planar process achieving 22.1M bits/mm^2 array density, providing only 100 μs *retention* time while using NWL to reduce transistor leakage. The achieved refresh power is 1.5 W/Gbit, which is 30% of the eDRAMs peak active power consumption. Another critical issue specifically for eDRAMs is that the leakage increases significantly at high temperatures, which is often the case for memory placed in close proximity to compute-intensive blocks such as CPUs/GPUs. For example in JEDEC DDR specifications [61], the refresh time interval (tRFEI) is reduced by 50% for operation above 85 °C due to increased leakage in bitcells. The impact of refresh on throughput can be up to 18% due to *read/write* traffic interruption caused by refresh commands and higher leakage at elevated temperatures.

In order to address the aforementioned DRAM design challenges, other than CMOS technologies have been explored. The Tunnel Field Effect Transistor (TFET) was proposed as a possible solution to reduce leakage while having the same scalability as MOSFETs. The conventional 1T1C DRAM architecture with TFET cannot work the same way as in CMOS because of the unidirectional current conduction. Therefore, there is a need to optimize the DRAM cell specifically for TFETs in order to utilize its advantageous properties over CMOS.

In [62] a capacitor-less TFET DRAM is shown using the potential well in a FDSOI-CMOS process. *Retention* times from hundreds of μs to ms are reported. However, predicting the stability is not possible with the present state of TFET device process variations and therefore, it is difficult to estimate the reliability of these circuits once fabricated.

The following sections present a refresh-free and scalable ultimate-DRAM (uDRAM) and its extension to SRAMs and CAMs for embedded applications. These memories are implemented with Si-TFETs and MIM capacitors using a 28 nm FDSOI-CMOS process, which allows co-fabrication of CMOS and TFETs.

4.2 The Ultimate-DRAM (uDRAM): TFET Negative-Differential-Resistance-Based 1T1C Refresh-Free DRAM

The uDRAM [63] relies on the *Negative Differential Resistance* (*NDR*) of the TFET observed in reverse bias as introduced in Chap. 2. The uDRAM bitcell utilizes the I_D as a function of reverse-biased V_{DS} characteristics of TFETs differing in operation from a CMOS DRAM cell; the 1T1C ultimate-DRAM cell consists of a TFET and a capacitor behaving as a static latch during *retention*. Figure 4.1a, b show the bitcell with the current and voltage setup during retention for storing a logical "0" and "1," respectively. In order to implement static-latch behavior during *retention*, the cell is designed to have I_{OFF} (device off-state current) $<<$ I_{CAPL} (leakage current in capacitor) $<<$ I_{NDR} (device current due to *NDR* property of TFET). The relation between currents and capacitor leakage is shown in Fig. 4.2 on the TFET

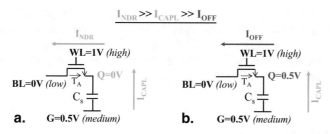

Fig. 4.1   TFET DRAM bitcell in retention, storing logical "0" and "1" [©2017 IEEE]

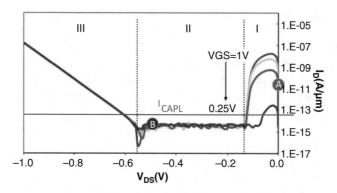

Fig. 4.2   TFET $I_D = f$ (reverse-biased V_{DS}) characteristics and capacitor leakage [©2017 IEEE]

$I_D = f(-V_{DS})$ device characteristics. The static latch behavior of the bitcell during *retention* is achieved with bitline (BL) low at 0 V, wordline (WL) high at 1 V, and the virtual ground node G connected through capacitor C_s to the storage node Q, at 0.5 V. With this setup the bitcell is biased in point A when storing a "0" with access transistor T_A operating in region I with low V_{DS} (0 V) and high V_{GS} (1 V), see Fig. 4.2. Two currents pull node Q in opposite directions, I_{CAPL} trying to charge up node Q while transistor T_A starting to discharge node Q due to I_{NDR} (region I) as soon as the voltage on $Q > 0$ V. Since $I_{NDR} >> I_{CAPL}$, 0 V representing logic "0" is maintained indefinitely in a static manner on Q. In the case the bitcell is storing a "1" the voltage across C_s is 0 V and T_A is in reverse bias in point B with $V_{DS} = -0.5$ V, see Fig. 4.2. Node Q will attempt to discharge through T_A due to I_{OFF} while at the same time I_{CAPL} keeps recharging node Q as soon as it goes below the node G voltage (i.e., 0.5 V). Since, $I_{CAPL} >> I_{OFF}$ (region II), the value of 0.5 V on node Q is maintained.

As the charge on the cell capacitor is maintained statically during *retention* the uDRAM presents the following advantages: (1) Significant improvement in terms of throughput and energy consumption due to refresh removal; (2) Unlike conventional CMOS DRAMs where capacitor scaling is limited by leakage and the higher *retention* time requirement, in the proposed design the minimum capacitor size is decided only by the bitline capacitance to be able to perform a *read*; (3) Unlike the high EOT of the capacitors in [55], which is used to reduce their leakage, the proposed design uses an EOT of 0.3 nm, which results in reduced area; (4) The smaller capacitance reduces the cost of the process used for eDRAM implemented within a single chip with CMOS logic.

4.2.1   Write Operation

Write operation of both "0" and "1" using a single TFET device in the conventional way of CMOS DRAMs is not possible due to the unidirectional V_G-controlled drain current in TFETs. In the proposed design, a *write* is performed by thermionic injection in the reverse-biased TFET (region III, Fig. 4.2) and using capacitive coupling between the virtual ground G and storage node Q. During *write*, all WLs are pulled down except for the one corresponding to the row to be written; BLs are pulled up or pulled down for writing data having a logical "1" or "0," respectively. In case of writing a "1," once the BL is pulled up the bitcell can be considered as initialized for *write* with the Q node rising to approximately 0.5 V regardless of the state stored before. This occurs due to the fact that with BL set to high voltage the access transistor T_A is in forward conduction and will pull up Q towards BL. After a short delay with respect to the BLs, a *write* pulse (ΔV) on node "G" is applied for the written row, see Fig. 4.3 for signal values. Due to capacitive coupling between G and Q, node Q will try to swing up by (ΔV) Volts on the rising edge of the pulse on G ($\simeq 1$ V for this example). Nodes Q in the selected row storing a "0" and having BLs at 0 V will be pulsed up to a value of 1 V. All other nodes in the row, i.e. nodes having BLs high and/or storing "1," start rising to (ΔV) + 0.5 V (1.5 V in this example).

Nodes Q in selected rows rising to more than 1 V with BLs at 0 V set access transistors (T_A) in thermionic injection regardless of gate voltage (V_{WL}), i.e. region III in Fig. 4.2. Therefore, node Q will discharge through T_A, reducing V_{DS} on T_A resulting in a limited rise in voltage. Discharge due to I_{OFF} is minimal in cells having BLs high in comparison to cells with BLs at 0 V. This results in a voltage difference between the nodes Q of selected cells written with "0" and "1."

Fig. 4.3 Signals during *write* [©2017 IEEE]

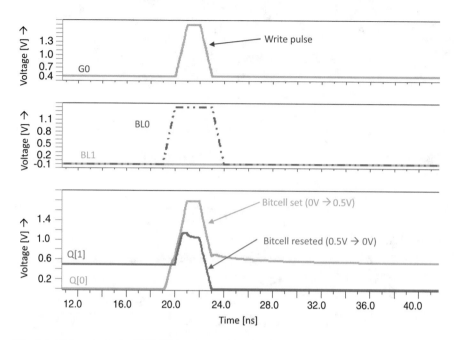

Fig. 4.4 *Write* waveforms [©2017 IEEE]

With the falling edge of the G pulse having the same ΔV of 1 V, node Q of the selected row is pulled down to either 0 V or 0.5 V depending whether cells have BLs at 0 V or 1.5 V, respectively.

Write waveforms for two bits in a selected row, Q[0] and Q[1], are shown in Fig. 4.4. A "1" is written in Q[0] initially storing a "0" by pulling up BL0 to 1.5 V during the G pulse ΔV; a "0" is written in Q[1] initially storing a "1" by keeping BL1 at 0 V. WLs and BLs are pulled to their *retention* voltages at the end of the *write* cycle. It should be noted that at the time when G is high the voltage difference between "1" and "0" can be tuned by adjusting the voltage on BLs during *write*.

4.2.2 Read Operation

At the start of a *read* BLs are precharged to 0.5 V and WLs are pulled down for all the rows except the one selected. Figure 4.5a, b shows *read* signal values for selected and partially-selected cells during *read*, respectively. Depending on the accessed cell value, "1" or "0," *read* BL either remains at the precharged value of 0.5 V or is discharged, respectively. Charge sharing between the BL capacitance and node capacitance C_s is determining the BL discharge value while reading a "0" from the bitcell. Figure 4.6 shows the *read* operation waveforms for "0" and "1." *Read* destroys the data in the cells storing a "0," therefore, *write-back* is needed at

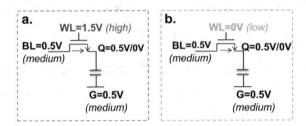

Fig. 4.5 Signals during *read*, (**a**) selected cell and (**b**) partially-selected cells due to precharged BLs [©2017 IEEE]

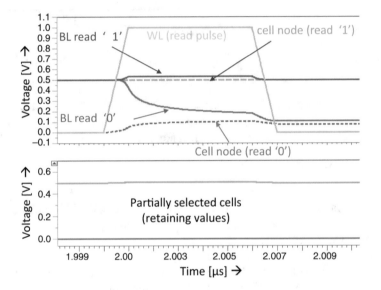

Fig. 4.6 *Read* waveforms [©2017 IEEE]

the end of the operation to restore the value in the cell, i.e., while closing the page as done in standard DRAMs.

During *read/write* operations the unselected rows of an array have storage nodes floating as access transistor (T_A) in the bitcells is OFF with a WL voltage at 0 V. Once the access is finished 0 V (logical "0") is retained/restored by the T_A, while the voltage of logical "1," i.e., 0.5 V, is always maintained by $I_{CAPL} \gg I_{OFF}$ as previously explained for cell *retention*. During access, even with floating Q for bitcells storing "0" in unselected rows, data is not disturbed because access time is too short to leak significantly the charge stored on C_s.

Fig. 4.7 2 × 2 Bitcell array organization [©2017 IEEE]

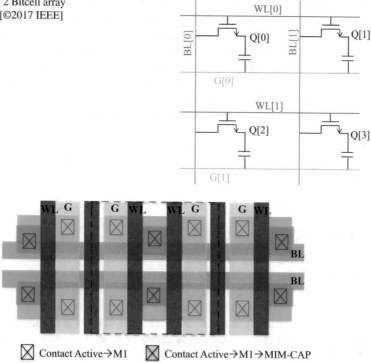

Fig. 4.8 uDRAM bitcell layout [©2017 IEEE]

4.2.3 uDRAM Bitcell Implementation and Performance

An uDRAM memory array organization is shown in Fig. 4.7. The memory is designed for implementation in a 28 nm FSDOI CMOS process with WLs in poly, MIM capacitors [55, 59, 60] and BLs in Metal-1 (M1), see Fig. 4.8. Unlike in [55] where high EOT is used for capacitor leakage reduction, EOT of 0.3 nm is used resulting in a smaller size. Bitcell capacitance C_s is 2.5 fF to match the BL capacitance for 128 cells. The design is implemented using planar transistors to ease fabrication with CMOS digital logic.

In CMOS DRAMs C_s is sized by the leakage through the access transistor/capacitor and *retention* time requirement. Due to the static nature of the latch in uDRAM, unlike in CMOS, the C_s requirement is relaxed and C_s equal to the array BL capacitance can be implemented in order to get up to 0.25 V BL discharge during *read*. This allows the value of C_s to be reduced significantly in comparison to conventional DRAMs and eDRAMs, by 70%–85% and 40%–60%, respectively. Leakage current in this design is <1 fA/bit on average assuming a 50% split between logic "1" and "0" storage in the memory, which represents more than two decades reduction in comparison to eDRAM [58]. Compared to DRAMs [3], leakage is reduced by up to 48× without using NWL and thick EOT for capacitor. Array dynamic power consumption during *write* is increased by 23% in comparison to standard CMOS DRAM mainly due to the switching of the virtual ground. However, the increase in dynamic power consumption is more than compensated by the energy gain due to refresh-free operation. Moreover, up to 18% throughput can be gained in comparison to standard CMOS DRAM due to refresh-free operation when used in a Dual-Data Rate (DDR) memory.

It should be noted that the explanation in this section uses specific voltages on signals as an example; these voltages can be tuned without any limitation to match the system requirements. The minimum requirement of the proposed uDRAM design is to have three supply voltages, i.e. low, medium, and high supply voltages, such as 0, 0.5 and 1.5 V.

4.3 uDRAM-Based 2T1C SRAM

Writing back the data in eDRAMs when replacing SRAMs close to compute-intensive blocks such as CPU, GPU or DSP is a major constraint. Therefore, eDRAMs are used as big-size L2 cache or lower in the memory hierarchy. In order to avoid a *write-back* during a *read*, another variant of the uDRAM, the ultimate-SRAM (uSRAM) [64], which works similarly to an SRAM is proposed. A 2T1C uSRAM implementation, which can be utilized to replace standard SRAMs to reduce area and leakage is shown in Fig. 4.9. In the proposed cell separate *read* and *write* ports, *read* bitline (RBL) and *write* bitline (WBL) are used to isolate the *read* from the charge stored on cell node Q with an extra transistor. Data *retention* and *write* operation are similar to a uDRAM.

The *read* operation is performed using RBL and WL. WLs are pulled down for the selected row during *read* with precharged RBLs (high). With WL pulled down, M_2 is switched ON or OFF depending on the value stored in the cell, "1" or "0," respectively. During *read*, RBL either remains at the precharged value of 1 V or

Fig. 4.9 2T1C uSRAM bitcell (retention bias voltages) [©2017 IEEE]

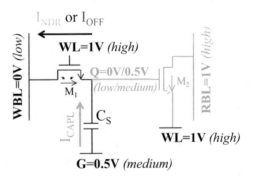

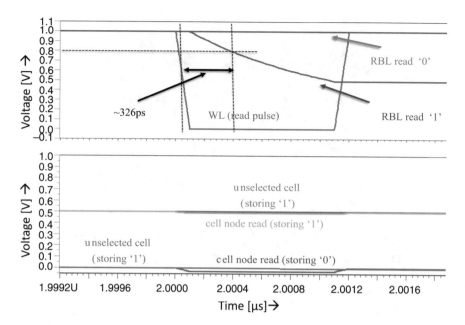

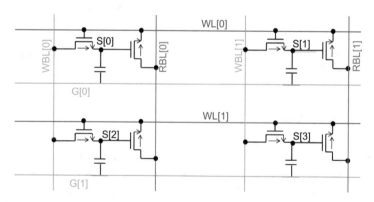

Fig. 4.10 *Read* waveforms showing two RBLs (read "0" and read "1") with active low WL. Internal cell nodes retain data during *read* [©2017 IEEE]

Fig. 4.11 2T1C uSRAM bitcell array organization [©2017 IEEE]

RBL discharges depending on the stored data, "0" or "1," respectively. Waveforms for reading "1" and "0" are shown in Fig. 4.10.

An uSRAM memory array organization is shown in Fig. 4.11. Bitcell implementation uses WLs in poly, MIM capacitors [55, 59, 60] and *read/write* BLs in Metal-2 (M2), see Fig. 4.12; similar to the uDRAM, the uSRAM is designed with a 2.5 fF bitcell capacitance C_s in a 28 nm FDSOI-CMOS process using planar transistors to ease integration with CMOS digital logic.

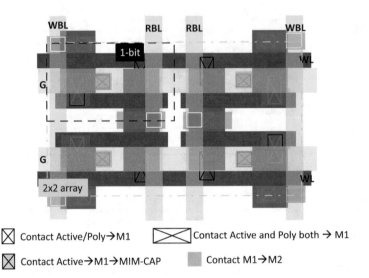

Contact Active/Poly→M1   Contact Active and Poly both → M1

Contact Active→M1→MIM-CAP   Contact M1→M2

Fig. 4.12 uSRAM bitcell layout

Fig. 4.13 3T1C uCAM
bitcell

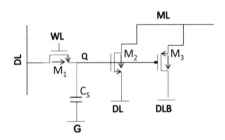

4.4 uDRAM-Based 3T1C CAM

A further extension of the uDRAM concept is the 3T1C ultimate-CAM (uCAM) cell. The resulting cell becomes an "extended" uSRAM cell with an additional PTFET *read* port to enable data comparison functionality required for CAM operation. As shown in Fig. 4.13 the *read* port consists of two TFETs with data lines DL and DLB providing search data and its complemented value, respectively. At the onset of CAM *search* the match line (ML) is precharged to the high voltage value and floated. ML will only discharge if either Q is "1" and search data is "0" (i.e., $DL = 0$ and $DLB = VDD$) or Q is "0" and search data is "1" (i.e., $DL = VDD, DLB = 0$), corresponding to a CAM miss. In case of a CAM hit, ML remains at the high (precharged) voltage value. Similar to the operation of a standard CAM at the end of a CAM *search* all the MLs are discharged towards GND except the one where a data match ("hit") occurred, which remains at the precharged voltage value. It should be mentioned that the application of the uDRAM core is particularly attractive for uCAM applications owing to the fact that CAM operation

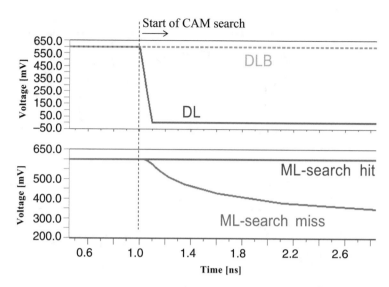

Fig. 4.14 uCAM *read*: hit and miss conditions

is vastly dominated by the *read* while *write* is very rare. As a consequence, the high energy cost of a *write* operation in the uDRAM bitcell, which is the factor limiting energy gains, is even-further mitigated in CAM-like applications in comparison to DDR-like applications. The *search* energy per bit of the uCAM cell can be assumed to be in a similar range as for the standard CAM, as the same number of signals are being triggered. The difference consists in shorter total metal lines for the uCAM due to a more compact cell and the fact that two transistors are connected to ML per cell for the uCAM. As a consequence, the uCAM can be assumed to use similar *search* energy as the standard CAM with significantly lower leakage and smaller area. *Read* waveforms for hit and miss conditions for different search values are shown in Fig. 4.14. Bitcell array organization and connections are drawn in Fig. 4.15. Figure 4.16 shows the layout of a 2 × 2 uCAM bitcell matrix using three metals with virtual ground G in the above metal layer.

4.5 Summary

The *NDR* property of TFETs and capacitor leakage are used to implement a refresh-free TFET DRAM bitcell, the uDRAM; memory architecture, operation and its physical implementation have been presented. The storage node capacitance (C_s) is reduced by 70%–85% and 40%–60% in comparison to conventional CMOS DRAMs and eDRAMs, respectively. The uDRAM bitcell area in 28 nm FDSOI CMOS is estimated at $0.0275\,\mu\text{m}^2$ representing the best-case area for planar technology. However, in a TFET implementation in order to insure the bitcell yield,

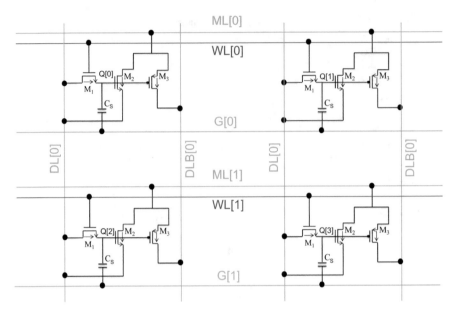

Fig. 4.15 uCAM 2x2 bitcell array

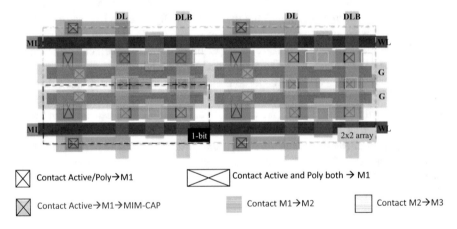

Fig. 4.16 uCAM bitcell layout

area may increase by 10–20%; area estimation with compact-memory design rules is $0.07\,\mu\text{m}^2$ and $0.104\,\mu\text{m}^2$ for uSRAM and uCAM, respectively. Further area shrinkage for all the proposed designs is possible by using a high-cost process to implement the access transistor vertically, as in the case of modern DRAMs [65]. The proposed uDRAM architecture is scalable and can be tuned to meet performance and area requirements by either adjusting the block sizes or supply voltages.

Table 4.1 uDRAM, uSRAM, and uCAM comparison with state-of-the-art CMOS RAM

	Proposed			State-of-the-art [58]	
	uDRAM	uSRAM	uCAM	eDRAM	LV SRAM
Technology (nm)	28	28	28	22	22
Bitcell area (μm^2)	0.0275	0.07	0.104	0.029	0.092
Refresh req.	No	No	No	Yes	No
Retention time (μs)	Inf	Inf	Inf	100	Inf
Capacitance/node (fF)	2.5	2.5	2.5	14.2	–
Leakage (fA/bit)	< 1	< 1	< 1	> 100	++
Sub-array latency (ns)	< 4	< 1	< 1	3	< 0.5

In a DDR configuration throughput gains of up to 18% are obtained owing to the removal of refresh, the achieved performance being compatible with DDR 1600 standard timing. The design summary and comparison with state-of-the-art is presented in Table 4.1. Far lower leakage for uDRAM, uSRAM, and uCAM is obtained in comparison to state-of-the-art 22 nm eDRAM and LVSRAM while maintaining similar performance, bitcell area, and array density despite the difference in technology node, see Table 4.1. Moreover, the up to 5.6× reduced value of the uDRAM bitcell capacitance allows a cheaper and more reliable process integration. The uSRAM bitcell area is 40% smaller than that of the 6T-CMOS high-density bitcell in 28 nm FDSOI allowing 1.4× more memory in the same footprint while reducing standby power by a few decades. The availability of uCAM provides an option to use a built-in CAM with uDRAM or uSRAM embedded on the same chip, using it as a cache to implement search functions in an area- and power-efficient way. uCAM is 47% better in area comparison with the high-density 6T-CMOS CAM based on a standard SRAM bitcell [66].

Chapter 5
TFET *NDR* Flip-Flop

5.1 Introduction

Cost and power efficiency are an important aspect for applications such as Internet-of-Things (IoT) and Wireless-Sensor Nodes (WSN). In SoCs optimized for these specification, key focus is put on SRAMs and flip-flops as they are the main contributors to area, energy, and leakage. Flip-flops in particular are critical components for synchronous logic and microprocessor-based systems where they are used as pipeline registers, register files, and data-buffers. These systems are often used in applications, which may run on energy scavenging/small batteries requiring low-voltage operation. In IoT applications a small form factor is important as even for a low- to medium-performance microprocessor more than 1000 flip-flops are required; therefore, optimizing area of flip-flops is an important consideration for IoT SoCs.

Currently the majority of digital systems are CMOS based, therefore, CMOS flip-flops are well explored for power and performance optimization [67–69]. Recent flip-flops implemented in other-than CMOS technologies, such as TFET flip-flops, [70], are designed similarly to the CMOS ones. These technologies could have advantages over CMOS; the TFET is one of the promising alternatives to explore for low-voltage and low-power flip-flop designs. However, due to the unidirectional property of TFETs, their higher drain capacitance and non-saturating $I_D - V_{DS}$ characteristics, standard flip-flop architectures face major performance constraints. The above-mentioned drawbacks limit the use of TFET-based flip-flops. This necessitates the investigation of alternative TFET circuit architectures in order to mitigate the pitfalls and take full advantage of the unique TFET properties for optimal flip-flop design.

This chapter investigates ultra-compact low-voltage flip-flop design using Si-TFETs for ULP applications requiring long battery life while providing good performance. A summary of the area and power efficiency of the existing CMOS and TFET circuit architectures is presented in Sect. 5.2.

© Springer Nature Switzerland AG 2021
N. Gupta et al., *TFET Integrated Circuits*,
https://doi.org/10.1007/978-3-030-55119-3_5

An ultra-compact flip-flop design using Si-TFETs for ULP application focusing on area and power efficiency using the *Negative Differential Resistance* (*NDR*) and unidirectionality properties of TFETs, see Chap. 2 [14], is presented in Sect. 5.3. The design-level issues in TFET flip-flops are analyzed and a novel 12T-TFET master-slave flip-flop (MSFF) cell designed using the *NDR* property of TFETs in reverse bias [14] is demonstrated. The used TFETs are compatible with CMOS for fabrication allowing the implementation of heterogeneous cores within a single FDSOI-CMOS process using both TFET and CMOS devices.

5.2 State-of-the-Art TFET Flip-Flops

Due to TFET's characteristics being different than those of CMOS, flip-flop architectures require additional modifications than the simple replacement of MOSFETs with TFETs. Transmission gate, master-slave, semi-dynamic, sense amplifier, and pseudo-static flip-flop architectures are presented in [70].

As shown in Fig. 5.1, the transmission-gate flip-flop [70] is designed using two transmission gates and two latches working as a master-slave flip-flop. A TFET transmission-gate flip-flop can be designed by directly replacing MOSFETs with TFETs while giving proper consideration to the direction of currents in transmission-gate P/NTFETs in order to have bidirectional currents. This circuit [70] has the following limitations: (1) the TFET's switch-off voltage is close to zero which could result in high leakage due to process variability; (2) it is not efficient due to the use of the NTFET in the transmission gate for charging the node and the PTFET for discharging; it would be more efficient if it was vice versa; (3) it takes longer to fully charge or discharge the circuit nodes in comparison to CMOS due to non-saturating $I_D - V_{DS}$ dependence in TFETs; (4) the maximum possible supply

Fig. 5.1 TFET transmission-gate flip-flop [©2013 IEEE]

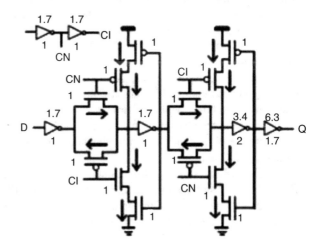

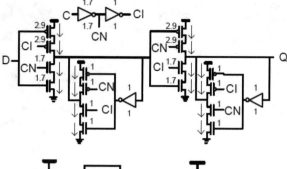

Fig. 5.2 TFET master-slave flip-flop [©2013 IEEE]

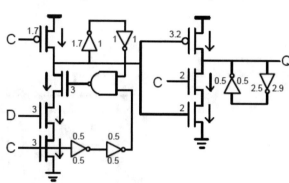

Fig. 5.3 TFET semi-dynamic flip-flop [©2013 IEEE]

voltage of operation is limited below 200–300 mV due to the high parasitic current (region III current for reverse-biased V_{DS}, see Chap. 2, Fig. 2.9) in transmission-gate hetero-junction TFETs. The operating voltage of homo-junction TFETs could be raised up to 600 mV.

Another master-slave flip-flop design using tri-state inverters is shown in Fig. 5.2 [70]. Tri-state inverters ensure that TFETs never operate with reverse-biased V_{DS}, thus allowing usage of this flip-flop at higher voltages. However, stacking of TFETs with non-saturating $I_D - V_{DS}$ dependence results in the significant degradation of performance, in particular under scaled supply voltage.

A semi-dynamic flip-flop design with dynamic precharge is shown in Fig. 5.3 [70]. The inverter chain propagation delay on the clock controlling the switch-off of the NTFET through the NAND gate defines the data capture window. This can result in a significant impact on timing because of a too small or too large capture window with process variation. Therefore, sizing poses a challenge and has to be done properly.

A sense-amplifier-based flip-flop is shown in Fig. 5.4 [70] using a differential sense amplifier to control the NAND gates of the SR-latch. A transmission-gate structure is used in place of a pass transistor due to the unidirectional behavior of TFETs. The flip-flop with modified latch is shown in Fig. 5.5. Limitations in such a flip-flop structure are similar to the other flip-flop structures using transmission gates, i.e., the supply voltage is limited by the parasitic current through transmission

Fig. 5.4 TFET sense
amplifier-based latch [©2013
IEEE]

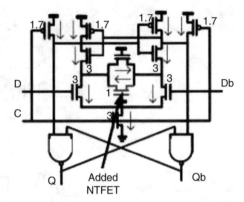

Fig. 5.5 TFET sense
amplifier-based flip-flop
[©2013 IEEE]

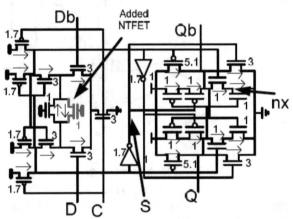

gates. In this architecture, the lower supply voltage impacts speed significantly due
to transistor stacking.

Figure 5.6a shows a MOSFET master-slave pseudo-static D Flip-flop design
[70], which could not be used directly with TFETs due to their unidirectional
property. Therefore, the circuit in [70] is modified for TFETs by implementing a
discharge path for nodes X and Y as shown in Fig. 5.6b. However, this circuit is
also limited to low-voltage operation due to parasitic currents through TFETs under
reverse-biased V_{DS}. This results in higher inverter delays and circuit timings (Clock
(C)-to-Q and setup/hold).

The speed comparison for all TFET architectures is also presented in [70]. The
comparison results show that some of the TFET flip-flops are better in terms of
setup, hold, and C-to-Q timing than CMOS. However, in order to maximize the
ON current the switch-off voltage of the TFET devices is close to 0 V. This results
in significantly higher leakage power consumption nullifying the fundamental
advantage of using TFETs in the first place. Variability can make the situation even
worse as some of the devices may be always ON having a negative OFF voltage and
resulting in even higher leakage. In order to make them usable, the OFF voltage

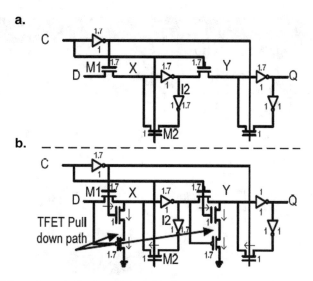

Fig. 5.6 TFET pseudo-static flip-flop [©2013 IEEE]

of TFET devices used in the flip-flop architectures described in [70] should be increased to a positive value for NTFET and negative value for PTFET at the cost of degraded performance of flip-flops.

5.3 12T-TFET Master-Slave Flip-Flop (MSFF) Design

We propose a new TFET-FF design [71] that does not suffer from the issues of the aforementioned designs using a 2T-TFET latch and a 2T-TFET tri-state inverter as shown in Fig. 5.7; the unidirectional property of TFETs allows implementation of the tri-state inverter by using only two transistors. The input (D) is connected to the input of tri-state inverter (I1) consisting of M0 (PTFET) and M1 (NTFET). When the clock (CLK) is low, M0 and M1 work as an inverter. When CLK is high, inverter I1 is in tri-state because M0 and M1 have reverse-biased V_{GS}. Similarly, the other inverter I2 (M4 and M5) is in tri-state when CLK is low and works as inverter when CLK is high. Master and slave latches are implemented using the *NDR* property of TFETs [14]. The operation of the *NDR*-based latch is similar to that of the SRAM latch shown in Fig. 3.19 with gate bias voltages replaced by CLK and CLKN signals in order to implement a tri-state latch. Output drivers are used to isolate the internal storage latch from the fan-out. The proposed design implements flip-flops with inverted output using 12-TFETs and with non-inverted output using 14-TFETs.

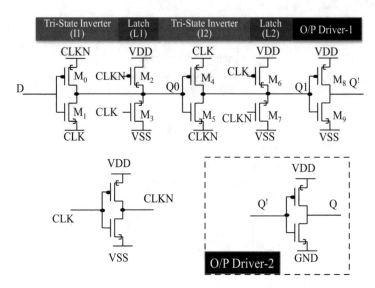

Fig. 5.7 Proposed 12T (14T with O/P driver-2) TFET-FF design [©2016 IEEE]

5.3.1 Flip-Flop Operation

The principle of operation of the flip-flop in Fig. 5.7 is as follows. When CLK is low D is complemented by inverter I1 (M0/M1) to Q0, which is the input to I2 (M4/M5); the master latch L1 (M2/M3) and inverter I2 (M4/M5) connected to Q1 are tri-state when CLK is low; therefore, the Q1 value is preserved by slave latch L2 (M6/M7) and outputs $Q^!$ and Q are driven by output drivers OP Driver 1 and 2. When CLK is high, input inverter I1 is in tri-state and master latch L1 is preserving the value on Q0 driving Q1 using inverter I2. During this period slave latch L2 is in tri-state. The waveforms on the flip-flop inputs, output, and internal nodes for data and clock transitions are shown in Fig. 5.8.

The 12T-TFET MSFF has the following advantages in comparison to conventional flip-flop designs: (1) at any given point in time, half of the TFETs in the flip-flop are in reverse bias condition (negative and positive V_{DS} for NTFETs and PTFETs, respectively) resulting in extremely low leakage for supply voltages up to 0.6 V (<3 fA/flip-flop); (2) there is neither feedback nor inverter delay within the latch, resulting in *setup time* (T_{Setup}) and *CLK-to-Q delay* (T_{CP2Q}) reduction, especially for low-voltage operation where inverter delays are large; (3) the total number of transistors is almost half of the conventional master-slave flip-flop resulting in reduced area. Due to the above-mentioned advantages, the performance of the 12T-TFET MSFF in terms of speed and power is better in comparison to CMOS, FinFET, and other TFET flip-flops reported in literature.

Fig. 5.8 Flip-flop voltage
waveforms, (**a**) Input data and
clock, (**b**) inverted output,
and (**c**) FF internal nodes

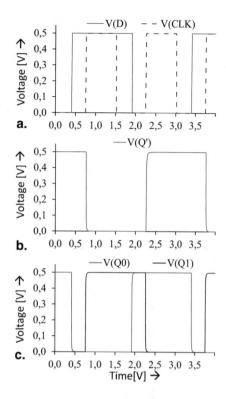

5.3.2   Energy Efficiency

In order to demonstrate the energy efficiency of the proposed TFET master-slave
flip-flop (MSFF) (Fig. 5.7) a C^2MOS MSFF similar to the one presented in Fig. 5.2
is designed in Bulk-CMOS and FinFET technologies with device size optimization
done in a similar way to industrial flip-flops. Figure 5.9 presents the leakage power
consumption for the proposed TFET design, CMOS, and FinFET (Low static power
(LSTP) and high-performance (HP)) MSFF implementations. The TFET MSFF
static power consumption is reduced by 4–7 orders of magnitude in comparison
to CMOS and FinFET designs. This is obtained owing to intrinsically ultra-low
leakage of the TFET and by adapting the architecture to avoid any parasitic current
path due to device unidirectionality at higher reverse-biased V_{DS}.

The dynamic power consumption for the three designs is shown in Fig. 5.10. The
flip-flop internal node capacitance of the proposed design is far less than that of
CMOS and FinFET flip-flops due to a lower number of devices and the low C_{GS} of
TFETs, see Fig. 2.10, Chap. 2; thus, dynamic power consumption for the proposed
design is 3–5 orders of magnitude better than that of CMOS and FinFET designs.

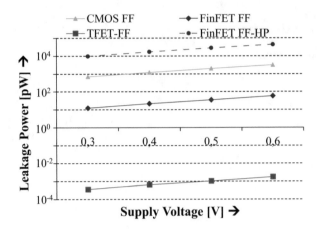

Fig. 5.9 12T-TFET FF comparison with LP-CMOS and FinFET FF's: leakage power vs. supply voltage [©2016 IEEE]

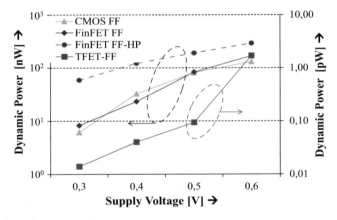

Fig. 5.10 12T-TFET FF comparison with LP-CMOS and FinFET FF's: dynamic power consumption vs. supply voltage [©2016 IEEE]

5.3.3   Performance

Usually, the flip-flop performance is evaluated in terms of T_{Setup} and T_{CP2Q}. Figure 5.11 shows the T_{Setup} requirement for the proposed TFET, CMOS, and FinFETs (LSTP and HP) MSFF designs. It should be noted that the proposed TFET design is faster than that of CMOS and FinFET-LSTP for a supply voltage range of 0.3–0.6 V. However, the setup requirement is almost similar to that of FinFET-HP, which is the fastest overall but also the most power consuming. At 0.3 V supply T_{Setup} of the TFET design is 14.6× and 56× lower in comparison to the CMOS and FinFET-LSTP designs, respectively. However, the T_{Setup} is still 3.75× larger than that of the FinFET-HP MSFF at 0.3 V supply voltage.

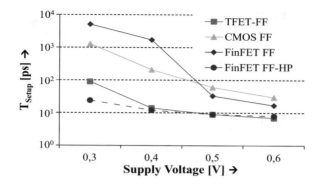

Fig. 5.11 12T-TFET FF comparison with LP-CMOS and FinFET FF's: T_{Setup} vs. supply voltage [©2016 IEEE]

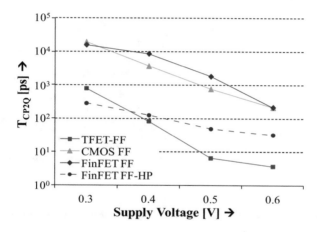

Fig. 5.12 12T-TFET FF comparison with LP-CMOS and FinFET FF's: T_{CP2Q} vs. supply voltage [©2016 IEEE]

Figure 5.12 shows the T_{CP2Q} for the implemented designs. Due to the lower number of delay elements in the CLK to Q path and lower capacitance, T_{CP2Q} is significantly reduced for the 12T-TEFT MSFF; T_{CP2Q} is 20× to 58× less than that in CMOS and FinFET-LSTP designs, respectively. For sub-0.4 V operation, T_{CP2Q} is 8× better for FinFET-HP at the cost of 10^7× more leakage power consumption in comparison to the proposed TFET flip-flop (Fig. 5.9).

Hold time for the proposed and the other two MSFF designs is mainly dependent on the delay of the inverter generating CLKN from CLK. Therefore, it is similar for all the implemented designs.

The maximum speed of operation of the flip-flop is limited by the T_{Setup} and T_{CP2Q}. Figure 5.13 shows the theoretical limit of the operating frequency defined as $1/(T_{Setup} + T_{CP2Q})$, vs. the supply voltage for the implemented designs. Up

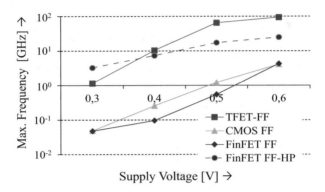

Fig. 5.13 12T-TFET FF comparison with LP-CMOS and FinFET FF's: maximum operating frequency vs. supply voltage [©2016 IEEE]

Table 5.1 Transistor count for various flip-flop designs [©2016 IEEE]

Flip-flop architecture	Transistor count
Proposed design	12 (14)[a]
Transmission gate	24
C^2MOS (C^2MOSFF)	24
Semi-dynamic (SDFF)	23
Sense amp (SAFF)	18 (19)
Modified sense amp (MSAFF)	26 (27)
Pseudo-static DFF (DFF)	14 (18)

[a]Device count will increase by two considering CLK buffering inside FF

to 23× improvement is achieved for the proposed TFET design in comparison to CMOS and FinFET-LSTP designs.

5.4   Comparison and Results

This section presents the comparison of the 12T-TFET MSFF with designs reported in literature for MOSFET and TFET flip-flops where the performance reported in [70] is used as a reference. Comparison is done for all designs under similar constraints such as fan-out and operating voltage. The different MSFF architectures considered in this comparison are summarized in Table 5.1 together with the corresponding transistor count. The performance data presented in the following subsections are extracted either from simulations or from literature for comparison purpose [70].

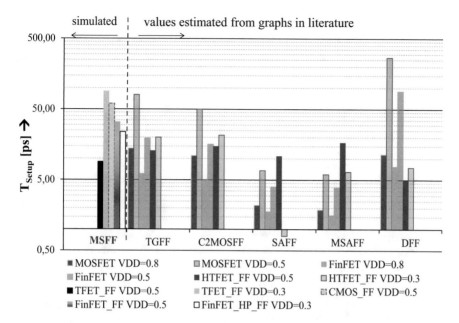

Fig. 5.14 Comparison—T_{Setup} for different flip-flop designs [©2016 IEEE]

5.4.1   Setup Time (T_{Setup})

Overall T_{Setup} of FinFET designs is better than all other TFET and CMOS flip-flop designs reported in literature. The T_{Setup} of the 12T-TFET MSFF is shown to be even better than that of the FinFET MSFF [70]. For the proposed design, T_{Setup} is achieved as similar or better to CMOS and FinFET flip-flops but with ultra-low leakage. Figure 5.14 shows the comparison of T_{Setup} for the proposed flip-flop design with various CMOS and FinFET designs.

5.4.2   Clock-to-Output Propagation Delay (T_{CP2Q})

Due to the optimized path from clock to output, T_{CP2Q} of the 12T-TFET MSFF is the lowest at 0.5 V supply voltage in comparison to all other flip-flop designs implemented with MOSFETs, TFETs, and FinFETs. T_{CP2Q} of the proposed design and the comparison with the other designs are shown in Fig. 5.15.

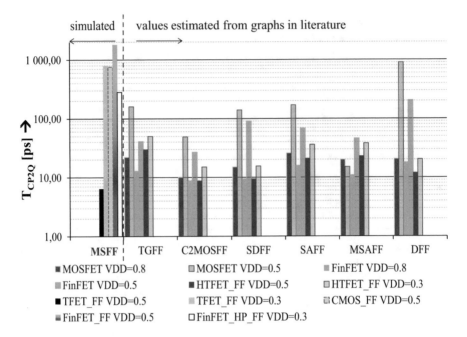

Fig. 5.15 Comparison—T_{CP2Q} for different designs [©2016 IEEE]

5.4.3   Minimum Clock Period Required $T_{Critical}$

$T_{Critical}$ is defined as $T_{Setup} + T_{CP2Q}$: this represents the theoretical limit for the minimum possible clock period for proper flip-flop operation. As shown in Fig. 5.16, the 12T-TFET MSFF has a $T_{Critical}$ similar to that of the FinFET and heterojunction-TFET flip-flops at 0.8 V and 0.5 V, respectively.

5.4.4   Leakage

The reduced leakage power in the proposed design in comparison to CMOS and FinFET FFs is due to the ultra-low OFF current (region II, see Chap. 2, Fig. 2.9) of the TFET ($I_{OFF} \approx 10^{-14}$ A/μm). Significant improvement in terms of leakage also in comparison to other TFET flip-flop designs [70] is due to an optimized TFET device for OFF current and the flip-flop architecture, which ensures TFET devices are never in a high parasitic current (region III, see Chap. 2, Fig. 2.9) operating condition by taking advantage of the *NDR* property. The focus for TFETs in [70] is to increase I_{ON} for better performance at the cost of leakage ($I_{OFF} \approx 10^{-9}$ A/μm). In our design more than half of the TFETs are always biased with reverse V_{DS} where I_{OFF} is at its minimum and is independent of V_{GS} operating

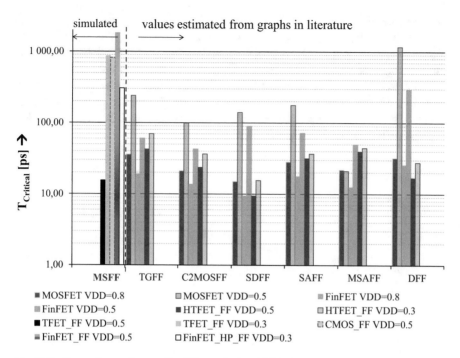

Fig. 5.16 Comparison—$T_{Critical}$ for different designs [©2016 IEEE]

in region II, see Chap. 2, Fig. 2.9. The enumerated features of our TFET device and proposed design result in the lowest static power consumption among all the implementations. In [70] only the leakage energy is reported, which is dependent on the timing window considered for measurement. Thus, an absolute comparison of static power consumption with [70] is not presented.

5.5 Summary

This chapter demonstrated the superiority of TFETs for implementing flip-flops; TFET's *NDR* and unidirectionality features were used to design a novel 12T-TFET MSFF architecture optimized for low cost and low power. The proposed flip-flop design uses 12/14 transistors (without clock buffered locally) and represents a promising architecture for near-threshold or subthreshold computing. The building blocks of the 12T-TFET flip-flop are the *NDR*-based 2T-TFET latch for data storage and 2T-TFET tri-state inverter taking advantage of unidirectionality and using CLK/CLKN as inverter supplies and latch biases. The flip-flop operates with ultra-low leakage current (<3 fA) for supply voltages up to 0.6 V. The TFET flip-flop supports voltage scaling and works from 0.6 V down to 0.3 V supply voltage.

The performance in terms of speed and power was evaluated and compared with published flip-flop designs using TFET, MOSFET, and FinFET devices. Leakage power reduction from 4 to 7 orders of magnitude compared to MOSFET and FinFET implementations was achieved. Dynamic power is reduced from 3 to 5 orders of magnitude in comparison to MOSFET and FinFET MSFF implementations. Maximum operating frequency for the proposed design is $23\times$ higher than MOSFET and LSTP-FinFET implementations but comparable to that of the HP-FinFET MS flip-flop while having $10^7\times$ less leakage.

Chapter 6
Content-Addressable Memories

6.1 Introduction

Content-addressable memories (CAMs) differentiate themselves from other memories as information is accessed by data instead of physical location. This makes CAMs popular in high-speed hardware-based *search* operations, such as look-up tables, data compression, image processing, register renaming [72], look-up buffers, and as highly associative caches in processors. However, using CAMs in systems efficiently is a challenge due to its high area cost and power dissipation. Since the parallel *search* operation is performed in the memory, optimization of speed and energy per *search* is crucial for an efficient CAM architecture. Another challenge for CAMs is the increasing leakage with every new technology node. Therefore, minimization of area and power consumption of CAMs for processor-based SoCs becomes an important design challenge.

In order to use CAMs as cache memory, it is important to optimize them for the specifics of each application. The best cache configuration depends mainly on the application requirements and design constraints leading to diverse cache architectures found in different processors. Therefore, it is difficult to find a single cache architecture fulfilling requirements of all applications. Examples include flexible architectures such as FPGAs with built-in ASIC blocks [73] or ARM cores with a large L2 (512 KB) cache, which needs a CAM for storing tags. ASIC and FPGA-based cache designs are proposed in [74–76], which can be configured dynamically to change the behavior in terms of associativity or mode of operation, cache or scratchpad, depending on the application requirements. However, while FPGA-based CAMs [76] have the advantage of flexibility, their performance and use-cases are limited. CAMs [74, 75] implemented in ASICs, on the other hand, are better optimized for area and speed but they have limited flexibility in comparison to the FPGA-based ones. Today, there is a need for designing flexible ASIC-implemented CAMs that can provide close-to-optimum performance for a large variety of applications.

© Springer Nature Switzerland AG 2021
N. Gupta et al., *TFET Integrated Circuits*,
https://doi.org/10.1007/978-3-030-55119-3_6

A possible solution is to design a Reconfigurable CAM/SRAM (ReCSAM), which can operate either as CAM and/or as SRAM depending on the application's needs. Sharing the memory array between CAM and SRAM optimizes the overall memory footprint by maximizing the average memory utilization and hence reduces total area and power while maintaining high speed of operation. For example, in applications which do not require a big L2 cache on a SoC as in [73], CAMs can be used as SRAM buffer or as register file. Sharing memory resources is an important aspect as the total memory capacity has to be increased to meet worst-case application requirements resulting in area overhead and leakage; however, few recent reports in literature focused on the reduction of CAM standby power [77, 78]. Techniques like fine-grain power gating, smaller *match* line, and pipelined *search* were implemented and benchmarked. CAM leakage power reported in literature is in the range of nA/bit [77] for CMOS technology. In order to further reduce leakage, other than CMOS technologies are also explored by researchers. In [78], the authors reported leakage in the range of pA/bit in CMOS with non-volatile elements (magnetic tunnel junctions) and hierarchal power gating.

Another aspect of CAM design is optimizing them for approximate-search, which allows to find the closest match of data in place of exact match. Such associative search in memories is useful for applications such as pattern search and face recognition.

This chapter analyzes the applicability of TFETs to content-addressable memories (CAMs) and presents ways to optimize the memory footprint in systems for LSTP applications using reconfigurability for longer battery life and/or energy harvesting.

In Sect. 6.2 we propose a TFET/CMOS hybrid reconfigurable CAM (ReCSAM), which can operate either as CAM and/or as SRAM depending on the application requirements [79]. TFET devices considered in this work are compatible with CMOS processes for fabrication. This allows the implementation of heterogeneous cores in a single FDSOI-CMOS process using both TFET and CMOS devices for arithmetic logic units and memories.

Also presented in this chapter, Sect. 6.3, is an extension of the TFET/CMOS CAM concept to a CMOS-only reconfigurable CAM/SRAM (ReCSAM) architecture based on the 6T-CMOS bitcell operating either as CAM and/or as SRAM depending on the application requirements [66] showing the advantages that can already be obtained on a mature technology (CMOS) having as a starting point the new design concepts inspired by an emerging technology (TFET) or by the hybridization of CMOS with TFET.

An extension of the ReCSAM to an associative memory architecture is proposed in Sect. 6.4. CMOS-only and TFET/CMOS hybrid architectures are proposed using 6T-CMOS and 8T-TFET bitcells, respectively. Section 6.5 describes Ternary CAM (TCAM) based on a novel TFET multi-bit latch using the *NDR* property of TFETs.

6.2 Hybrid TFET Reconfigurable CAM/SRAM Array Based on a 9T-TFET Bitcell

The Reconfigurable 9T-TFET CAM/SRAM (ReCSAM) design is meant for efficient memory utilization in a wide variety of applications. The ReCSAM cell with dual wordline is shown in Fig. 6.1 and the overview of the memory architecture is in Fig. 6.2. In order to optimize the overall memory leakage current, the bitcell array is designed with TFETs and the periphery with MOSFETs. This choice is based on the higher CMOS drive strength in comparison to that of TFETs for the same speed of operation but reduced periphery area at the cost of increased leakage. During the standby period the periphery can be power gated to minimize the leakage power consumption of the CMOS logic. Similar to the TFET-based 8T bitcell DPSRAM presented in Chap. 3, single-ended sense amplifiers (SA) are used to limit the bitline discharge, therefore, reducing power consumption and allowing a bigger column size. Several reports on single-ended sensing have been published in literature for CMOS, [51, 52], which can be used directly or with modification in the proposed design. A detailed description of single-ended SAs can be found in Chap. 7.

In the proposed design data words are stored vertically in a column aligned with bitlines and orthogonal to wordlines, see Fig. 6.2. When the memory operates in CAM mode data search is performed using the left bitline BLL as *match* line and wordlines WL1 and WL2 as *search* data lines. In SRAM mode the *read* operation uses BLR as wordline for selecting a column and *read*-bitlines RBLs as data lines for reading data bits of the word. The *write* operation is identical for both CAM and SRAM modes. Write is performed using bitlines BLL and BLR, and wordlines WL1 and WL2, see Sect. 6.2.1 for detail; all cells in one column are written simultaneously with the other columns held in *retention* mode. In a ReCSAM architecture the cell array is optimized and dynamic reconfigurability is

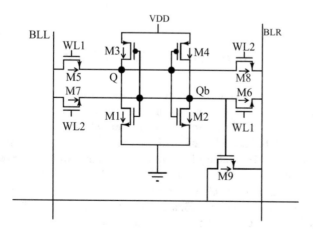

Fig. 6.1 9T-TFET CAM cell [©2016 IEEE]

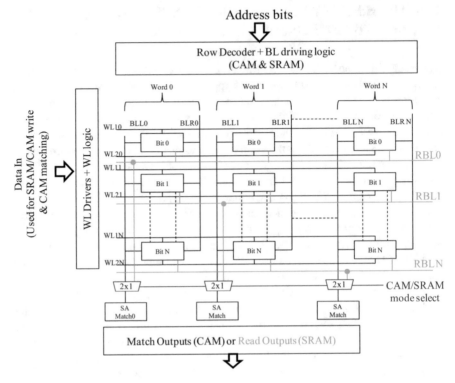

Fig. 6.2 ReCSAM architecture [©2016 IEEE]

added with additional digital logic in wordline and bitline drivers. Details of *write* and *read* operations during CAM and SRAM modes are presented in the following section.

6.2.1 *Write Operation (CAM and SRAM Modes)*

During *write* BLL is at VDD, BLR is at GND, and for every bit of the word, either WL1 or WL2 is selected depending on the value to be written in cell is "1" or "0," respectively. Except for the column to be written, all other columns are in *half-selection* (*HS*) with bitlines BLL and BLR placed at *retention*-mode voltage (0.6 V) to improve stability. The value of the *retention* voltage was chosen to minimize leakage exploiting the unidirectional behavior of TFETs as already explained in Chap. 2. Table 6.1 shows the voltages on BLs and WLs for selected and half-selected columns during *write* operations. Figure 6.3 plots the waveform of three complete *write* cycles demonstrating the cell's correct operation.

Table 6.1 Signal voltages during CAM-/SRAM-mode *write* for selected and *HS* cells (writing)

	Column write-1 (active cells)	Column write-0 (active cells)	Other columns (HS cells)
BLL	1 V	1 V	0.6 V
BLR	0 V	0 V	0.6 V
WL1	1 V	0 V	1/0 V
WL2	0 V	1 V	0/1 V
VDD	1 V	1 V	1 V

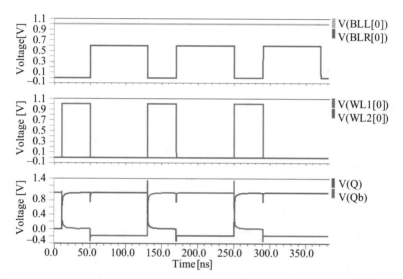

Fig. 6.3 CAM *write* operation [©2016 IEEE]

6.2.2 Read Operation (CAM and SRAM Modes)

All columns are read simultaneously during a CAM *search* to detect a hit or miss. By selective wordline activation per row the bitline BLL of each column works as wired logic AND to compute a hit or miss for each word. Sensing circuits are placed on BLLs at the bottom of the columns to detect a CAM hit or miss during the *search* and to pass the result on to the output. BLR is held at the retention voltage (0.6 V) during a CAM *read*, see Table 6.2. The CAM *search* is performed by activating WL1 for "1"s and WL2 for "0"s corresponding to the word to be matched by connecting each BLL across each row (two wordlines per row) to either Q or Qb of the cell, to be checked if the bitcell stores the same value as the corresponding bit in the *search* word. BLL is connected to Q for a search of "1" and to Qb for a "0" resulting in reading a "1" from that cell if there is a match, therefore, not affecting the precharged BLL (high), see Fig. 6.1. With this logic of WL selection, we read all "1"s in a hit and at least one "0" in a miss condition. All columns in a block

Table 6.2 Signal voltages during CAM- and SRAM-mode *read* [©2016 IEEE]

	CAM mode		SRAM mode	
	Active cells (search data = "1")	Active cells (search data = "0")	Cell read (active column)	Unselected columns
BLL (precharge voltage)	1 V	1 V	1 V	1 V
RBL (precharge voltage)	1 V	1 V	1 V	1 V
BLR	0.6 V	0.6 V	0 V	0.6 V
WL1	1 V	0 V	0 V	0 V
WL2	0 V	1 V	0 V	0 V
VDD	1 V	1 V	1 V	1 V

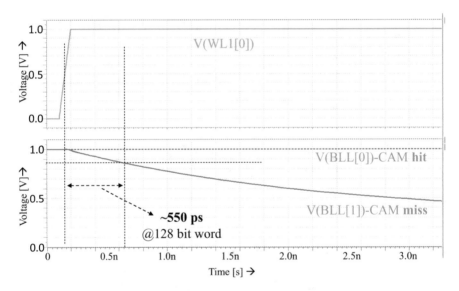

Fig. 6.4 CAM *search* operation, word[0]-CAM hit (BLL0), and word[1]-worst-case CAM miss (BLL1) [©2016 IEEE]

of a CAM can be read concurrently as matching of the words happens in parallel. Figure 6.4 presents *read* waveforms for CAM hit and miss cases for worst-case *read* speed, i.e., single-bit mismatch.

The SRAM-mode *read* is performed using RBL as bitline and BLR as active-low wordline; BLL, WL1 and WL2 are kept at *retention* voltages during SRAM *read*. Table 6.2 lists the voltages of different signals during CAM- and SRAM-mode *search/read* operation.

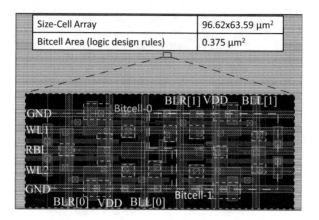

Size-Cell Array	96.62x63.59 μm²
Bitcell Area (logic design rules)	0.375 μm²

Fig. 6.5 Cell array and dual-bitcell layouts; VDD, BLR, and BLL in Metal-2; GND, WL1, WL2, and RBL in Metal-3 [©2016 IEEE]

6.2.3   Implementation, Results, and Comparison

Cell array and bitcell implementation are shown in Fig. 6.5. Wiring parasitics extracted from the layout are included in calculation of power and speed for the designed memory. The sizing of wordline (WL) drivers is done to optimize leakage while considering WL capacitances.

6.2.3.1   Performance

Read speed is dependent on bitline capacitances, i.e. BLL for CAM- and RBL for SRAM-mode *read*, respectively. Therefore, CAM-mode *read* speed is dependent on column size, while SRAM-mode *read* speed is dependent on row size. *Read* is analyzed for different column and row sizes for CAM and SRAM modes, respectively. *Write* speed is fairly independent of the column and row sizes, and depends on the voltages and current drive of transistors.

Read and *write* speed for CAM and SRAM operation are evaluated and compared in order to find the optimum column and row size for both modes. *Read/write* performance with assist techniques is also considered in this analysis. Since the proposed design uses single-ended *read* with BLR at *retention* voltage, WL boosting (WL1 or WL2) can be used as a *read-assist (RA)* technique to speed up the bitline discharge without impacting stability of bitcells in CAM mode. Negative-bitline *write-assist (WA)* technique can be used up to −150 mV on BLR without *write-disturb (WD)* issue because of the 150 mV V_{OFF} voltage of TFET devices.

WL boosting of 100 mV is considered as *RA* for CAM mode and negative bitline of −100 mV is used as *WA* for CAM-/SRAM-mode *write* operations. Figure 6.6 shows the minimum wordline pulse width (WLP_{MIN}) requirement for *read* and

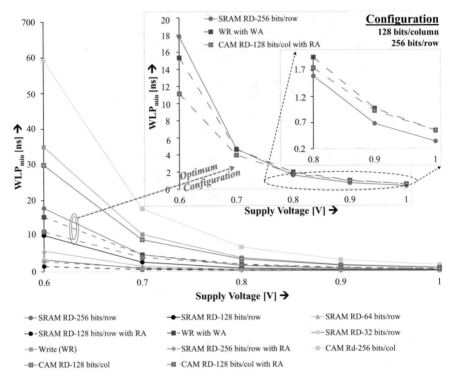

Fig. 6.6 WLP_{MIN} for *read/write* operations for CAM and SRAM modes vs. supply voltage [©2016 IEEE]

write operations with and without assist techniques for various column and row sizes. It should be noted that the optimum configurations are 256 cells/row with 128 cells/column and 128 cells/row with 128 cells/column, where *read/write* speed for CAM and SRAM modes are similar while using assist techniques for CAM/SRAM *write* and CAM *read* operations. Equal values for the column and row size allow full reconfigurability as the same IO logic is required for both CAM and SRAM modes. It should be noted that a row size of 256 is also compatible with a column size of 128; however, for this configuration half of the sense amplifiers in IO will not be used during SRAM mode.

6.2.3.2   Power Consumption

The power consumption of the designed memory and comparison with other implementations are presented in Table 6.3. Standby mode I_{LEAK} is computed with the periphery OFF and cell array power ON in order to retain the data. E_{READ} is defined as the energy consumed during *read* on row drivers and bitlines. Bitline discharge is limited to 200 mV and 100 mV for 9T-TFET ReCSAM single-

Table 6.3 TFET CAM/SRAM speed and power comparison [©2016 IEEE]

Cell	VDD (V)	WL_{Pulse} read (ns)	WL_{Pulse} write (ns)	E_{READ} (fJ/bit)	E_{WRITE} (fJ/bit)	I_{LEAK} active (without Rd/Wr) (pA/bit)	I_{LEAK} STBY (fA/bit)	Area-bitcell (μm^2)
9T-TFET (128 × 128) [CAM/SRAM]	1	0.87/0.19	0.92/0.92	0.14/2.38	16.5/16.5	25.5/3.19	5.00	0.375
9T-TFET (128 × 128) [CAM/SRAM]-Assist	1	0.55/0.19	0.55/0.55	0.16/2.38	16.5/16.5	25.5/3.19	5.00	0.375
TCAM(128 × 32) 65 nm [77]	1	0.24	N.A.	1230 uW[b]	742 uW[b]	N.A.	$> 1 \times 10^6$	
MFCAM(1 Mb) 90 nm [78]	1.2	5	5	25.7mW[b]			$3.95 \times 10^6 / 4.6 \times 10^3$	4.563
6T-CMOS 32 nm	0.75	0.27	0.16	9.10	12.9	11.8	7.8×10^3	0.120
6T-CMOS 65 nm [8]	1.2	7 ns (access time)[a]	N.A.	N.A.	N.A.	N.A.	27.0^a	2.04

[a]Measurement value for full memory
[b]Reported power per access

ended sensing and 6T-SRAM differential *read*, respectively. E_{WRITE} is the energy consumed during *write* operation. I_{LEAK} in active mode is the total leakage in the bitcell array and periphery operating as an SRAM with dynamic power gating with only 25% of the drivers active.

In estimating E_{READ} for CAM mode full array parallel search is considered, while for SRAM mode, single-word *read* is considered. Therefore, energy per bit is 17× lower for CAM mode in comparison to SRAM mode. E_{WRITE} is estimated considering single-word *write*, i.e., for a single column. Additional power-saving techniques are used for IO logic; for SRAM mode only 16 drivers, i.e. eight rows, are active with others power gated, while CAM mode needs all drivers active. Therefore, active-mode leakage is up to 87% less for SRAM mode in comparison to CAM mode. For comparison purposes, reports from literature with low-leakage Ternary CAM (TCAM), CAM, and SRAMs are also shown in Table 6.3. It can be seen that the standby power of the presented design is $10^6×$ less in comparison to other SRAM, CAM, and TCAM cells. The cell size is 3.12× larger than the standard 6T-CMOS SRAM cell but 5.4× smaller than the ultra-low leakage SRAM cell [8]; CAM/TCAM cells reported in literature are even bigger in size. Evaluated static noise margins at 1 V supply are 120 mV and 200 mV for *write* and *read*, respectively.

6.2.4   Summary

An ultra-low leakage reconfigurable TFET CAM/SRAM (ReCSAM), which can work as CAM and/or as SRAM has been presented in this section. The design is useful in order to efficiently utilize the available embedded memory for a wide range of applications. Less than 5fA/bit memory array leakage current is achieved at 1 V supply voltage, an improvement of up to $10^6×$ compared with state-of-the-art CMOS SRAM and CAM bitcells. The proposed CAM architecture supports voltage scaling and allows application of performance boosting techniques without impacting cell leakage. The minimum *write* access pulse for CAM and SRAM modes is evaluated at 1.37 ns at 1 V supply voltage; for *read* the evaluated access pulse is 1.39 ns and 1.03 ns at 1 V for CAM and SRAM modes, respectively. The design supports different configurations, such as variable tag size, variable word size and can operate as a combination of CAM and SRAM. These configurations can be chosen at implementation time or can be made programmable to dynamically adjust them as required during runtime.

6.3   ReConfigurable CAM Extension to CMOS

The CMOS ReCSAM bitcell is designed using similar concepts as the 9T-TFET ReCSAM bitcell presented in Sect. 6.2 such as dual wordlines and the bitline used as *match* line. The proposed CMOS ReCSAM design is meant for efficient memory utilization in a wide variety of applications. Reconfigurability allows runtime change between SRAM and CAM modes of operation. However, CMOS design challenges are different due to the bidirectional current flow between source and drain. Therefore, a new kind of sensing method is proposed, which allows operation of the design at high speed while maintaining reliability. The proposed memory supports single-cycle *read/write* operations and parallel search in all columns.

6.3.1 ReCSAM Architecture

The CMOS ReCSAM architecture is shown in Fig. 6.7, CAM data words are stored vertically in a column aligned with bitlines and orthogonal to wordlines. In CAM mode, a single-bitline configuration is used by shorting BL and BLB in the IO logic using a configuration bit (cam_mode). In this architecture, the cell array is optimized and dynamic reconfigurability is introduced with additional digital logic implemented in wordline and bitline drivers. The bitcell is using the sizing of the 28 nm FDSOI foundry library bitcell with modifications in Metal-2 and Metal-3 connections to include dual wordlines and vertical ground routing. The area of the implemented bitcell is $0.277 \, \mu m^2$ with logic design rules; with compact-memory rules the area would be reduced to $0.197 \, \mu m^2$. The different operation modes of this architecture are described in the following subsections.

6.3.1.1 Binary CAM (BCAM)

Data search is performed with the bitlines pulled up using the virtual ground "vs" of the bitcell working as *match* line and the wordlines wl1 and wl2 as *search* data lines. This approach improves the stability of the bitcell, reducing energy consumption and

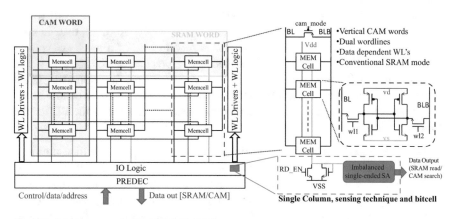

Fig. 6.7 Memory organization [©2017 IEEE]

Fig. 6.8 Single-ended imbalanced sense amplifier, precharged to "0" [©2017 IEEE]

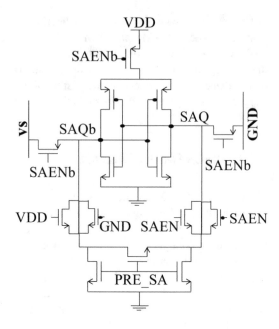

sharing the sense amplifier for both SRAM and CAM in read mode. All "1"s from the column are read in case of a hit and at least one "0" in case of a miss. This is achieved by setting the wordlines for all columns according to the search data, i.e., wl1 high for searching "1" and wl2 high for searching "0." The virtual ground "vs" of each column works as a logic-NAND to compute hit or miss for each word. In case of a miss at least one "0" in the bitcell is connected to the bitline, continuous current flows from the bitline to the bitcell ground "vs." Therefore, "vs" starts rising due to this current flowing through the diode-connected MOSFET during *read*, RD_EN = "0" (see Fig. 6.7 lower-right inset), ultimately leading to a reduction of the *read* current.

A single-ended imbalanced SA shown in Fig. 6.8 is connected to "vs," which is connected to VSS before the *read* with RD_EN = "1." A "0" is read in case of a hit, i.e., $V(vs) \approx V(VSS)$, and a "1" is read in case of a miss, i.e. voltage $V(vs) > V(VSS)$. The SA is implemented with a precharged voltage of 0 V and an imbalance of 100 mV. The imbalance is created by charge injection using MOS capacitors, as shown in Fig. 6.8. Waveforms for reading a hit and miss condition are shown in Fig. 6.9. The SA is implemented to sense "vs" instead of the BL for the following reasons:

- In the presented architecture bitcells are shorted through the bitline during a CAM *search*; the BL discharges almost to 0 V when more "0"s are connected resulting in a data loss for cells having a "1." In order to avoid this scenario either the bitcell pull-up should be stronger resulting in area penalty or *read-assist (RA)* such as WL lowering should be used, which results in a much slower *read* operation and adds design complexity. By using sensing on "vs"

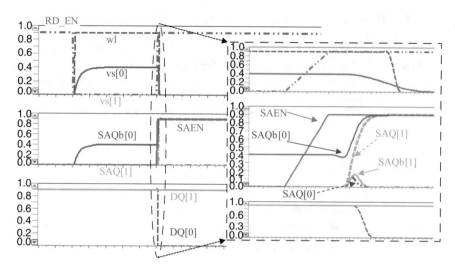

Fig. 6.9 Simulation waveforms of single-ended *read* during CAM *search* [©2017 IEEE]

the BL precharge is kept ON during *read* ensuring a limited voltage drop on BLs allowing bitcell sizing similar to that of an SRAM without stability issues during CAM *search*.

- With rising "vs," the *read* current decreases resulting in energy savings for the worst-cases, such as multiple-bits misses in a single column.

Exploiting "vs" for *read* has one drawback, which has to be taken into consideration when designing the "vs" footer circuit. The problem consists in the necessity to limit the maximum voltage that can appear on "vs" as it impacts directly the effective voltage across the SRAM bitcells and thus may impact cell stability. This is however easily obtainable even for multiple cells activated simultaneously due to the reduction of single-cell current contribution as "vs" increases and adapting the footer size to the maximum number of cells activated in the single operation in the column.

MOS capacitors used for creating the imbalance in SA are implemented as multiple devices, both PMOS and NMOS, to minimize cross-corner variation (Slow–Fast and Fast–Slow). Imbalance Tuning (IT) in SA can be done on chip to adjust for variation by selectively disabling the SAEN signal going to MOSCAPs by using configuration bits.

For *write* in CAM mode, wordlines work as data lines, i.e. wl1 = "1" for writing "0" and wl2 = "1" for writing "1," and both bitlines are pulled down for a selected word column.

6.3.1.2 Pseudo-Ternary CAM (TCAM) Mode

The pseudo-TCAM Mode differs from the binary mode in the following aspect: in a CAM *search* both wordlines can be left low for a row in order to mask the bits of that row in a *search* operation. This can be used to implement a global-mask operation, which is used in routing applications. The global mask can be implemented as a configuration register or it can be obtained from another SRAM storage [80].

6.3.1.3 SRAM Mode

In SRAM mode a *read* is performed using "vs" with just wl1 high for a single row, working as address selection for *read*. The *write* operation in SRAM mode is performed in the conventional way using bitlines as complementary data lines and both wordlines high for a single row acting as address selection.

6.3.2 Measured Results

The proposed memory architecture was fabricated as an 8 Kb (128x64 bitcell array) test-macro in a 28 nm FDSOI-CMOS process. The test-chip photo with the block layout details is shown in Fig. 6.10.

The *read/write/search* test patterns generated externally allow the extraction of the minimum wordline pulse duration (WLPulse_min) for stable operation and low-power consumption. The bitcells and memory architecture were characterized for stability during *read* and *write*. Figure 6.11 shows *read* and *write* WLPulse_min for different supply voltages. Application of imbalance tuning (IT) results in a

Fig. 6.10 Chip photo and test-macro layout [©2017 IEEE]

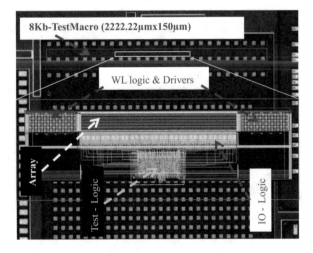

8Kb-TestMacro (2222.22µmx150µm)

WL logic & Drivers

Array

Test - Logic

IO - Logic

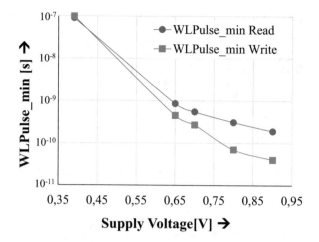

Fig. 6.11 *Read/write* performance [©2017 IEEE]

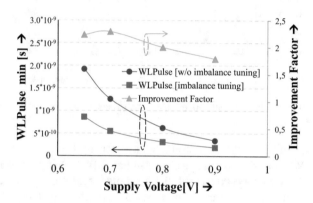

Fig. 6.12 *Read* speed with/without SA imbalance tuning [©2017 IEEE]

WLPulse_min of 190 ps at 0.9 V. The impact of IT on SA is shown in Fig. 6.12, giving a speed improvement of up to 56% at sub-0.9 V.

The distribution of the measured minimum bitcell supply voltage (VDDmin) for ten chips is shown in Fig. 6.13a. The WLPulse_min at the lowest functional voltage of 0.38 V was measured at 99 ns. The measured leakage/bit varies from 23.76 pA at 0.9 V to 4.35 pA at 0.4 V, see Fig. 6.13b. Pulled-up bitlines during CAM mode improve the *read* data stability. In the worst case, one cell storing a "1" and 63 a "0," are connected to bitlines; the bitcell data are stable for supply voltages ranging from 0.4 V to 0.9 V or above. The voltage increase on node "vs" during *read* is limited due to the *read* current reduction with increasing V(vs). In measurements this results in stable retention of data during *read* even for the worst-case scenario defined above.

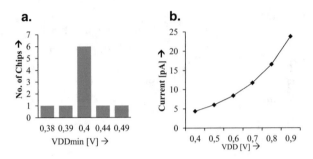

Fig. 6.13 VDDmin distribution for ten chips and bitcell leakage. (**a**) VDDmin distribution. (**b**) Bitcell Leakage [©2017 IEEE]

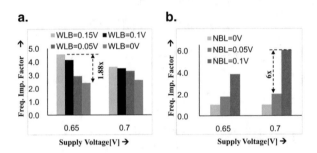

Fig. 6.14 Assist techniques analysis, (**a**) WL boosting, (**b**) negative bitline [©2017 IEEE]

The analysis of the impact of assist techniques for sub-0.8 V, i.e., negative bitline for *write* and WL boosting for *read* is shown in Fig. 6.14. Speed improvements of up to 1.88× and 6× were measured at sub-0.7 V supply for *read* and *write*, respectively. Table 6.4 presents the comparison of measured results with the state-of-the-art showing the following energy improvement figures: 4.6× [6], 8.3× [81], 5.9× [82], and 14.3× [83]. *Search* speed is also improved compared to the corresponding references by 4.2× [6], 3.12× [82], 6.24× [83].

6.3.3 Summary

A high-speed 6T-CMOS ReCSAM (Reconfigurable CAM/SRAM) memory architecture with a new energy-efficient sensing technique is proposed through extension of the TFET ReCSAM design introduced in Sect. 6.2. It is shown that starting from a new architecture (i.e., TFET/CMOS) and by redesigning the array, the new principles of design can be applied to the mature CMOS technology and important benefits can be obtained, see Table 6.4. A test-macro of 8 Kb was implemented in 28 nm FDSOI CMOS reaching up to 1.56 GHz at 0.9 V with 0.13 fJ/bit energy consumption per *search*, achieving an improvement of 4.6×

Table 6.4 CMOS CAM/SRAM comparison with state-of-the-art [©2017 IEEE]

	This work	[6]	[81]	[82]	[83]
Technology	28 nm FDSOI	28 nm FDSOI	32 nm	65 nm	0.13 μm
Transistors/cell	6T	6T	11T	10T	9T + Read
Area/cell (μm²)	0.197 μm² [a]	0.152 μm²	–	3.3	20
Array size	128 × 64	64 × 64	(64 × 64) × 4	128 × 128	128 × 32
Frequency (VDD)	1.56 GHz (0.9 V)[b] 8.90 MHz (0.38 V)[c]	370 MHz (1 V)		500 MHz (1 V)	250 MHz (1 V)
Energy/search/bit (fJ)	0.13 (0.9 V)	0.6 (1 V) 0.41 (0.75V)	1.07 (1 V) 0.3 (0.5 V)	0.77 (1.2V)	1.87 (1 V)
Match-line technique	1-Single-ended imbalanced SA	2-Single-ended SA	Wide AND	NOR	Differential
Memory modes	BCAM/SRAM/ Pseudo-TCAM	BCAM/TCAM/ SRAM	BCAM	BCAM	BCAM

[a] Area with compact design rules (with waiver on metal routing)
[b] Calculated value = 300 ps periphery delay estimation + measured WL_{MIN}
[c] Calculated value assuming measured WL_{MIN} is 80% of the clock cycle at 0.38 V

to 14.3× over published CAM/SRAMs. The *search* speed is also improved up to 3.12× to 6.24×. The implementation is compatible with compact 6T-SRAM foundry bitcells.

6.4   Associative Memory Architecture

This section presents the design and operation of an associative memory architecture that can be used in search applications such as pattern matching and neuro-inspired computing. In particular, this section focuses on the operation of an associative memory, which can be implemented using CMOS, FinFET, TFET, or non-volatile memory technologies. Reports on this subject in literature describe associative memory implementations either in software for algorithms or emulations in neuromorphic networks. Most of the research done on hardware implementations of associative memories is oriented towards neuromorphic computing. Neuromorphic implementations are used for applications such as pattern matching [84, 85] and approximate computing for power-performance trade-off; examples of the latter are approximate Floating-Point Unit (FPU) or Special-Purpose Units (SFUs) using associative memory [86–90]. Associative memories such as ReRAMs and STTRAMs reported in literature are implemented using non-volatile memories. Non-volatile memory technologies are limited in usage because of the lack of maturity and compatibility with CMOS in terms of operating voltages and speed. Software implementations of associative memories are used in algorithms with hashing methods for lookups, e.g. IP lookups, longest prefix matching, etc. However, software implementations are compute intensive and cannot be used in application-specific hardware where area/power needs to be optimized. Associative memory implementation using CMOS memories provides the best option for using it as embedded IP in ASICs. This section focuses on extending the SRAM/CAM presented in the previous section to associative memory architectures.

6.4.1   Associative Memory Architecture with CMOS CAM
Bitcell

In this section we propose a CAM-based low-power and area-efficient associative memory architecture. As shown in Fig. 6.15 the fundamental idea is to use optimized comparison logic on *match* lines of the CAM to find the closest match to the word. The optimized comparison logic to be integrated with the CMOS CAM is based on Winner-Take-All (WTA) logic. Figure 6.16 shows a 6T-NOR CAM cell compatible with this architecture. The bitlines shown on both sides of the cell are shorted in WTA and used as a single bitline (BL) per column similar to the CMOS ReCSAM cell introduced in Sect. 6.3, Fig. 6.7. In the proposed architecture words are stored

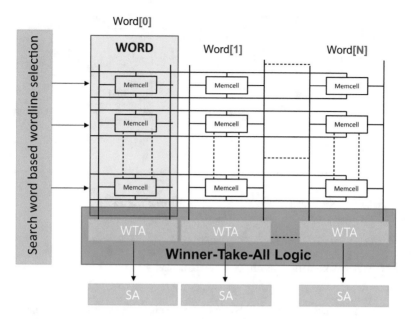

Fig. 6.15   Associative memory architecture

Fig. 6.16   6T-CAM bitcell

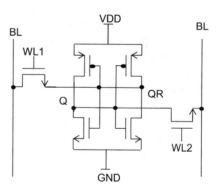

vertically in a column, bitlines are used as *match* lines and WTA comparison logic is used to compare all *match* lines in order to find the closest match. The comparison provides one-cold encoded output, i.e., only one output is logic-low and all others logic-high, which is followed by sense amplifiers to resolve fully and provide digital outputs.

Two-bit WTA-based comparison logic is shown in Fig. 6.17. Its operation is based on sinking the maximum BL discharge current for the closest match. In order to achieve maximum discharge current for the closest match, the search-data bits are inverted before comparing; this is in contrast to normal CAM-mode operation as presented in Sect. 6.3. For the cell shown in Fig. 6.16, during *search* mode WL1 or WL2 is selected depending whether the data to be looked for is "1" or "0,"

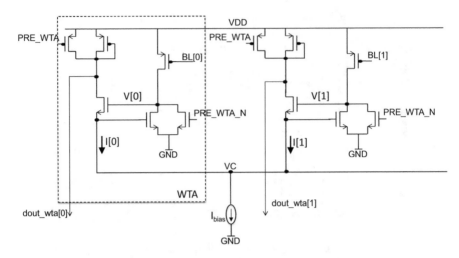

Fig. 6.17 Winner-take-all implementation for two columns

respectively; in case of a match BL will be discharged through either node QR for cell storing a "1" or node Q for cell storing a "0," respectively; BL discharges through a single or multiple bitcells depending on the number of bits matching. In case of a mismatch, the bitcell node storing a "1" is connected to BL; therefore, there is no BL discharge through mismatched bits. In order to find the closest match the maximum BL discharge current through bitcells needs to be detected with the bitline pull-up switched ON. The BL voltage is decided by the ratio of the BL pull-up current and BL discharge current through the bitcells of matched bits. The WTA logic compares the BL voltages and provides one-cold output identifying the closest matching word. The *write* operation is done in a similar way as explained for the 6T-ReCSAM, i.e. by pulling down the BL for the word/column to be written and selecting WL2 or WL1 for rows depending on the data to be written, "0" or "1," respectively.

6.4.1.1 Associative Search Operation

During a *search* to detect a hit or a miss condition per bit, the objective is to read a "0" from all cells (bits), which are matching the searched data bits. WL1 or WL2 is selected for each row depending on the bits of the matching word, see Fig. 6.16. This is done by connecting the BL of the column to the node of the cell, QR or Q, expected to store a "0" for a match as explained above. With this logic of WL selection we read "0" (i.e., bitline discharges) for a hit and "1" (i.e., the bitline does not discharge) for a miss. All columns in the CAM array are read concurrently for matching the words in parallel. Table 6.5 lists the signal voltages during the CAM *read/search* operation. Standard *read-assist* techniques, such as wordline under

Table 6.5 Signal voltages during *Search*

	Active cells (search data =="0")	Active cells (search data =="1")
BL (pull-up ON)	1.0 V	1.0 V
WL1	0 V	1.0 V
WL2	1.0 V	0 V
VDD	1.0 V	1.0 V
GND	0.5 V (VDD/2)	0.5 V (VDD/2)

drive *read-assist* (WL under drive-RA), can be used to improve the *read* stability of the cell.

During *search* the highest BL discharge current, i.e. the lowest BL voltage, is present on the word/column, which has data closest to the search data, i.e. the highest number of "0"s read. The WTA logic can be implemented to compare *read* currents on BL or the corresponding BL voltages. Figure 6.17 shows a voltage-comparison WTA circuit, where the BL voltage for different columns is compared to find the lowest voltage. The BL voltage is decided by the ratio of pull-up current to the total BL discharge current of BLs in the word/column. As shown in Fig. 6.17 BL voltages are compared by the WTA logic to generate a one-cold encoded dout_wta[*] output.

The circuit operation is described as follows:

I_{bias} sets the maximum current through the WTA logic of the entire array. For each different BL voltage, each WTA, i.e. per BL, will intend to consume current. By considering, for instance, the case in Fig. 6.17 with BL[0] going lower than BL[1], V[0] goes higher than V[1] meaning the corresponding pull-down n-type device is more conductive for column 0 than for column 1. As a result, the NMOS on the left in column 0 has a higher V_{GS} than the corresponding NMOS in column 1 and I[0] is higher than I[1] due to "vc" being set by I_{bias}; the corresponding output dout_wta[0] will consume almost all I_{bias} leaving other branches with almost zero current. This results in dout_wta outputs going high for all the columns except the one conducting the I_{bias} current. Figure 6.18 shows the WTA logic for each memory bank, one per column, and a common current source for the bank. Waveforms for column 0, 1, and 2 are shown during *search* in Fig. 6.19 with 4, 3, and 1 bit mismatches, respectively.

6.4.1.2 Write Operation

The *write* operation is the same as described in Sect. 6.3. During *write* BL of the selected column/word is at "GND" and for every bit of the word either WL1 or WL2 is selected depending on the value to be written into the cell, "1" or "0," respectively. Except for the column to be written, all other columns are in *HS* mode with bitlines (BL's) placed at *retention*-mode voltage ($VDD/2$) and the virtual ground of bitcells at $VDD/2$. This is done to reduce power consumption in *HS* columns even when

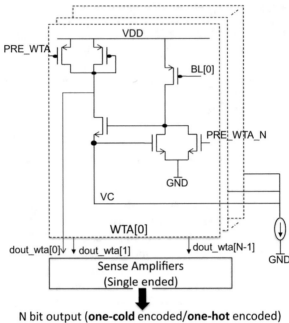

Fig. 6.18 Winner-take-all logic for one memory bank; one WTA per column

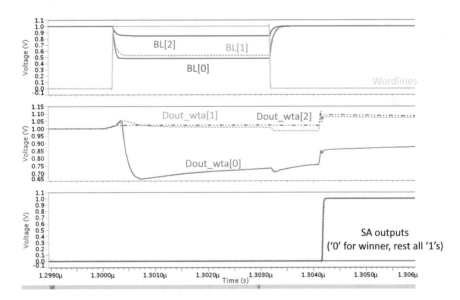

Fig. 6.19 *Search* operation waveform, simulation condition (mismatch): Col[0]-4 bit, Col[1]-3 bit, Col[2]-1 bit

Table 6.6 Voltages during *write* for selected and *HS* cells (writing "1")

	Column written (active cells)	Other columns (HS cells)
BL	0 V	0.5 V
WL1	1 V	1 V
WL2	0 V	0 V
VDD	1 V	1 V

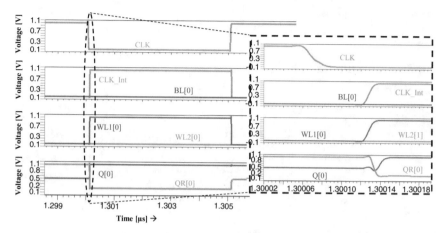

Fig. 6.20 *Write* operation with bitcell ground at $VDD/2$

WLs are activated. Voltage values on the different lines during *write* are listed in Table 6.6 with waveforms plotted in Fig. 6.20.

6.4.2   Alternate TFET and CMOS Architectures

The implementation of the described associative memory architecture can be done in different technologies, such as CMOS, Tunnel FETs, ReRAMs, and with a different kind of comparison logic, such as current-mode WTA. A few of the alternate designs are described below.

6.4.2.1   Dual-Port Associative Memory Architecture Using 8T-TFET DPSRAM Bitcell

This section presents an SRAM-based low-power and area-efficient dual-port associative memory architecture. The fundamental idea is similar to the CAM presented in Sect. 6.4.1 to use optimized comparison logic on *match* lines of the CAM to find the closest match to the searched word, see Fig. 6.15. The difference in this architecture, see Fig. 6.21, is that two words, Word[0] and Word[1], are stored

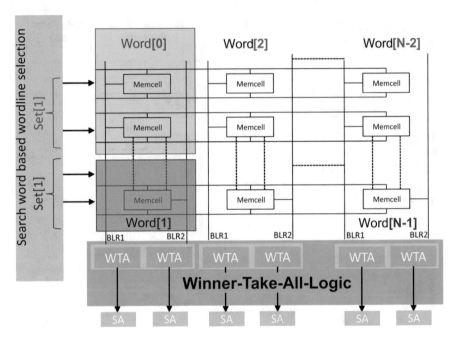

Fig. 6.21 TFET associative memory architecture

Fig. 6.22 8T-TFET associative memory compatible bitcell

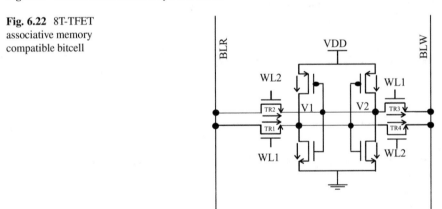

in a single column. This is done utilizing the fact that due to the unidirectional TFET conduction, it is possible to design longer columns w.r.t. CMOS without impacting the cell stability. Figure 6.22 shows a compatible 8T-cell, which is similar to that of the TFET SRAM introduced in Sect. 3.3.

In this architecture words are stored vertically in a column, the bitline (BLR's) in all columns are split in two bitlines (BLR1 and BLR2); half of the bitcells in column are connected to BLR1 and other half to BLR2. However, there is only one BLW for the column connected to all cells (not shown in Fig. 6.21). This allows in a single column to perform two parallel *search* operations and one- or two-word *write* at any given time. BLRs are used as *match* lines with active pull-ups. WTA comparison logic is used to compare all *match* lines in order to find the closest match. Comparison provides one-cold encoded output, which is followed by sense amplifiers to fully resolve the matching and provide digital outputs. The same WTA comparison logic shown in Fig. 6.17 can be used for TFET memories as well. In order to achieve the maximum discharge current on BLR for the closest match, Fig. 6.21, it is necessary to get the maximum number of zeroes connected to BLR during a *match*, see the detailed explanation in Sect. 6.4.1 above. In case of mismatch, the bitcell node storing a "1" is connected to bitline BLR, therefore, there is no BLR discharge through mismatched bits. In case of a match BLR is discharged either through node V1 or node V2 depending on the selected wordline. In case of a partial match BLR discharges through a single or multiple bitcells depending on the number of matching bits. In order to find the closest match, the maximum BLR discharge current through bitcells with bitline pull-up switched ON needs to be detected. The BLR voltage will be decided by the ratio of the resistance of BLR pull-up and BLR discharge path through bitcells for matched bits. BLR1 and BLR2 are used to *search* simultaneously for two words, one from the upper half of the column and the other from the lower half (see Fig. 6.21), respectively. The two BLRs work identically as described above, each with its own WTA logic block.

The *write* operation is performed in a similar way to the SRAM, except that the BLW for the selected word, i.e. one column, is activated by pulling it low and selecting WL's according to the data. Two-word *write* is possible by activating WL's for both sets of words and pulling BLW low.

6.4.2.2 Associative Memory Using Standard 10T-NOR CAM Bitcell

The proposed associative memory can also be implemented in CMOS using a standard 10T-NOR CAM bitcell shown in Fig. 6.23. Similar comparison logic as the one shown in Fig. 6.17, is implemented on *match* lines with *match* precharge ON during *search* operation. The basic architecture is shown in Fig. 6.24 and overall memory organization is shown in Fig. 6.25. *Write* is performed using bit and bit_b lines as data lines with the wordline acting as word selection line. The *match* operation is done using bit and bit_b as *search* data with output on the *match* line (ML).

6.4.2.3 Current-Mode WTA Logic

The WTA logic for this memory can be implemented using current comparison. Figure 6.26 shows an example of a standard WTA circuit where input currents I_1 to I_N represent BL *read* currents mirrored in the WTA circuit. An application for current-mode WTA is an associative memory using 10T-NOR CAM bitcells as shown in Fig. 6.27.

Fig. 6.23 10T-CMOS CAM
cell

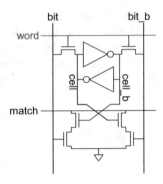

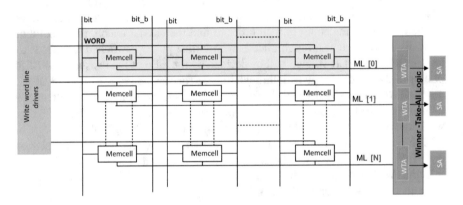

Fig. 6.24 Bitcell array of 10T-NOR CAM-based associative memory

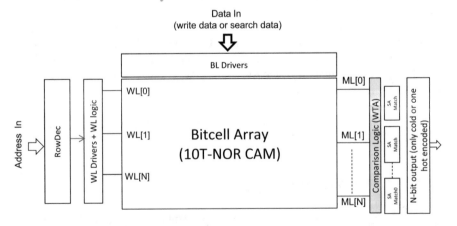

Fig. 6.25 Overall memory organization with WTA logic on *match* lines

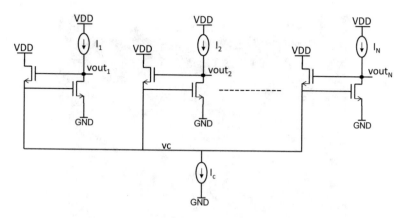

Fig. 6.26 Current-mode WTA circuit

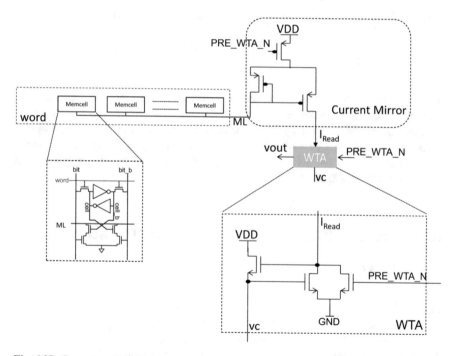

Fig. 6.27 Current-mode WTA-based comparison logic for associative memories

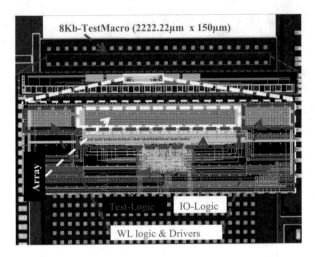

Fig. 6.28 Associative memory test-chip photo and block layout description

6.4.3   Summary

The memory architectures presented in Sect. 6.4 are applicable for use-cases implementing approximate-*search* operations such as pattern search, automatic language detection, face recognition, implementation of approximate floating-point units in particular for GPUs or video processing. Figure 6.28 shows the test-macro of an 8 Kb 6T-CMOS CAM-cell-based associative memory architecture fabricated and operational in a 28 nm FDSOI process.

6.5   Ultra-Low-Power TFET Ternary CAM (TCAM)

6.5.1   TCAM Cell

The proposed 7T-TCAM bitcell is shown in Fig. 6.29; TFETs M_0, M_1, and M_2 are biased such that they are always operating at a reverse V_{DS}, i.e. $V_{DS} < 0$ for NTFETs and $V_{DS} > 0$ for PTFETs. Due to the *NDR* property the TFETs M_0, M_1, and M_2 maintain the $V_{DS} \cong= 0$ condition owing to the hump current $0 \leqslant |V_{DS}| \leqslant 0.1$ V, see Fig. 2.5, Chap. 2. This results in three valid storage combinations where Q0Q1 can be "00," "10," or "11." For a supply voltage of $VDD = 0.6$ V, "00" is maintained by M_1 and M_2 with both having $V_{DS} = 0$ V and M_0 having $V_{DS} = 0.6$ V, i.e. a minimum current; similarly "11" is maintained by M_0 and M_1 with $V_{DS} = 0$ V and M_2 with $V_{DS} = -0.6$ V; "10" is maintained by M_0 and M_2 with $V_{DS} = 0$ V, and M_1 with $V_{DS} = 0.6$ V. The value of "01" is an unstable condition, i.e. a forbidden state, with M_1 switched ON in forward

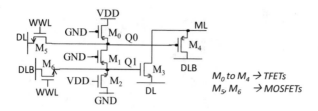

Fig. 6.29 Hybrid TFET/CMOS 7T-TCAM bitcell [©2016 IEEE]

V_{DS} condition. The proposed latch, therefore, can store ternary data according to the above-mentioned three combinations of data. "0" is represented with "00" on Q0Q1, "1" is represented by "11," and "X" or "don't care" by "10." M_1 needs to conduct the hump current with $V_{GS} = 0$ V in order to make "0" and "1" stable at nodes Q0 and Q1, respectively. Another option for devices where the hump current is fully suppressed at $V_{GS} = 0$ V, is to use either negative gate voltage on M_1 or add one more NTFET device in parallel between nodes Q0 and Q1 with the gate at VDD and V_{DS} negative.

The *read* port consists of two TFETs, M_3 and M_4, with data lines DL and DLB providing the *search* data and its complemented value, respectively; WWL is low and M_5 and M_6 are OFF during *read*. The *match* line ML only discharges if either Q0Q1 is "11" and search data is "0," i.e. DL = 0 and DLB = VDD, or Q0Q1 = "00" and search data is "1," i.e. DL = VDD and DLB = 0. For data match and "don't care" combinations "10" on Q0Q1, ML remains precharged. The *read* waveform is shown in Fig. 6.30 for hit and miss conditions for different search values. Device and wiring parasitics are estimated and included in *read* speed evaluation.

The *write* operation is performed using the *write* wordline (WWL), NMOS transistors M_5 and M_6, and data lines DL and DLB. Since $I_D(M_5, M_6) \gg I_D(M_0, M_1, M_2)$ (i.e., hump current), it results in fast writing of node Q0 and Q1. The *write* operation for different combinations of values is shown in Fig. 6.31. Wordline boosting is used to overcome the limitation of using an NMOS for writing "1" on nodes Q0 and Q1.

The memory array organization is shown in Fig. 6.32; VDD, GND, DL, and DLB are routed vertically, while WWL, ML are routed horizontally. The bitcell area is optimized by reducing the number of transistors and metal lines required per bit in comparison to a conventional 16T-TCAM bitcell. For a minimum footprint a 14T dual-bitcell layout block is designed to have an even number of TFETs with an area of 0.293 μm^2/bit, see Fig. 6.33.

6.5.2 Summary

A multi-bit latch using the TFET *NDR* property is proposed with a possible application for Ternary CAMs. The latch works for a voltage range from 0.3 V

Fig. 6.30 *Read* waveform for hit and miss conditions [©2016 IEEE]

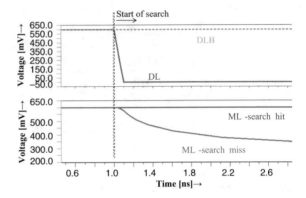

Fig. 6.31 *Write* waveform for writing X (10), 0 (00), and 1 (11) [©2016 IEEE]

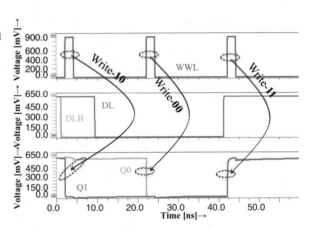

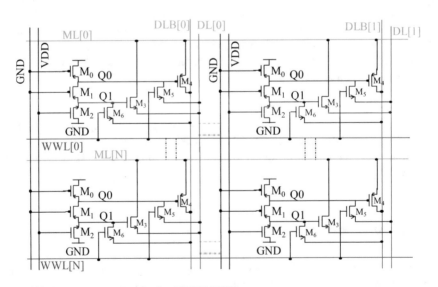

Fig. 6.32 Memory array organization [©2016 IEEE]

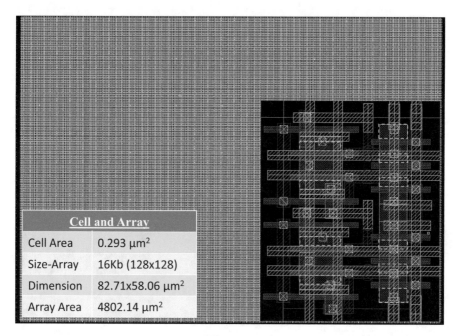

Cell and Array	
Cell Area	0.293 μm²
Size-Array	16Kb (128x128)
Dimension	82.71x58.06 μm²
Array Area	4802.14 μm²

Fig. 6.33 TCAM dual-bitcell layout [©2016 IEEE]

to 0.6 V. A TFET/CMOS hybrid 7T-TCAM bitcell for a compact TCAM using a single ternary TFET latch is proposed. The ultra-compact TCAM is designed by optimizing the ternary latch and removing separate storage of the mask bit. This results in more than 50% reduction in the number of transistors in the bitcell in comparison to a conventional CMOS 16T-TCAM. Two bitlines/data-lines are reduced in comparison to the conventional CMOS TCAM cell by removing storage of an independent mask bit. The proposed cell has ultra-low standby power and works for a voltage range from 0.4 V to 0.6 V for the bitcell array. The standby power is improved by more than two decades in comparison to state-of-the-art low-leakage TCAMs [78]. The achieved leakage is less than 2 fW/bit with sub-ns *read* and *write* delays at 0.6 V supply for a word size of 256.

Chapter 7
Sensing Techniques

7.1 Introduction

Conventional differential sensing is used in most CMOS memories; single-ended sensing has already been reported in literature and used in CMOS memories but is limited to specific use-cases such as 8T-CMOS SRAM, CMOS-DRAM, Flash, non-volatile memories [51–53]. Single-ended sensing is a promising option to use for optimized TFET memory cells due to the unidirectionality property of TFETs, which represents an obstacle for providing a differential output to the SA. Therefore, most state-of-the-art TFET memories are using single-ended sensing for *read*. Most TFET memory bitcells presented in literature have static power consumption several decades below that of their CMOS counterparts but exhibit limited performance. Therefore, the main challenge for designing single-ended sensing for TFET memories is to reliably differentiate "1" and "0" while limiting the required bitline voltage drop with a compact Sense Amplifier (SA), as for many applications the SA has to fit in the column pitch. Especially for compact memories, such as the 3T-TFET bitcell memory presented in Chap. 3, either a standard SA with tall and inefficient layout or an inverter-based SA can be used to meet the column pitch.

In the case of single-ended sensing using a differential SA a voltage source is needed to generate the reference voltage while for inverter-based sensing the bitline needs to be discharged below the trip point of the inverter, which often results in a full discharge, leading to a slow operation and higher dynamic power consumption for *read*. To alleviate these problems and improve speed an inverter-based sense keeper can be implemented in the read circuit [52].

Another aspect capturing the attention of researchers are adaptive memory techniques. Adaptive techniques are useful in various ways for increasing the reliability at low voltages and minimizing design time margins leading to speed and power improvement. These techniques are reported in literature in SRAM designs adaptive in terms of architecture. A lot of research is done on memories toward the

© Springer Nature Switzerland AG 2021
N. Gupta et al., *TFET Integrated Circuits*,
https://doi.org/10.1007/978-3-030-55119-3_7

analysis of performance, area, and power constraints; various architectures were proposed, including new kind of memory cells, flexible ECC implementations, and run-time configurable-memory architectures [91–94]. The most interesting proposals are reviewed below.

The design presented in [91] uses cells of different sizes in the same memory in order to optimize area vs. error rate. MSBs are stored in larger cells and LSBs are stored in smaller cells. In [92], the authors proposed a flexible architecture where *write assist (WA)* and ECC can be enabled partially whenever needed. The number of ECC bits can also be increased to afford more soft errors in MSBs with the ECC bits stored in LSBs. In [93], an adaptive memory architecture is presented using dual imbalanced sense amplifiers in place of a single balanced SA. In [94], an adaptive *write* wordline (WWL) pulse-width modulation scheme is described. *Write* completion is monitored every cycle; when *write* completion is detected, a Built-In-Self-Test (BIST) stops the *write* operation; with this, the presented architecture supports adaptive WWL voltage-level modulation to work as *WA*. The WWL is boosted after the first cycle in order to assist the *write* operation while half-selected bitcells are not disturbed due to bitline regeneration during the cycle. Since the majority of the memory area, especially for SRAMs, is occupied by bitcells, it is important to minimize bitcell area. Therefore, for adaptive memories it is important that the *sense* and *write* circuits implement this feature without impacting conventional bitcell design and sizing.

The first part of this chapter focuses on different single-ended techniques useful for TFET and/or CMOS memories. In the latter part adaptive sensing techniques and their application in memory architectures are presented. The sensing techniques presented in this chapter are introduced for use in TFET-CMOS hybrid memories. However, these techniques are equally useful in TFET-only memories, and most of them, which are not using any TFET-specific properties are also applicable for CMOS-only memories.

7.2 Charge-Injection-Based Single-Ended Imbalanced Sense Amplifier

This section presents single-ended sensing using an imbalanced differential SA based on charge injection described in [50]. An imbalanced SA is shown in Fig. 7.1. It can be noticed that one input serves as reference connected at VDD and the other input, the sensing node, is connected at the bitline to be read, BLR. Therefore, with a balanced single-ended SA it is not possible to read a "1" using VDD precharged bitlines as in the case of a differential SA because BLR is at VDD and the other SA node, its reference, is also precharged at VDD. The design principle of an imbalanced SA uses charge injection to create an imbalance for removing the reference voltage source and uses symmetric sizing of devices in order to minimize variations. The *read* scheme uses an SA imbalanced by 100 mV relative to BLR,

Fig. 7.1 Imbalanced sense
amplifier circuit diagram
[©2015 IEEE]

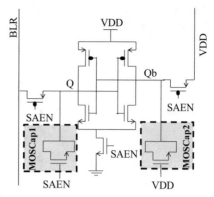

Fig. 7.2 Waveform: *Read*
"0" using imbalanced sense
amplifier [©2015 IEEE]

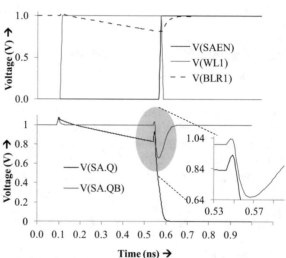

see Fig. 7.1. In the case of reading a "0," the bitline discharge should be more than
100 mV in order to be read correctly. For the case of reading a "1" BLR remains
at the precharged value (VDD) and the SA will resolve with node Q kept at "1." In
order to ensure correct functionality taking into account parameter variations, the
SA is designed with a 3-sigma offset of 50 mV. Therefore, to ensure correct *read*
operation BLR should discharge by 150 mV for reading a "0." For reading a "1,"
BLR should discharge less than 50 mV including noise, leakage, and variations. As
shown in Fig. 7.1, the imbalance is designed with charge injection using MOSFET
capacitors. MOSCap1 is sized to inject a charge generating a 100 mV impulse on
SA internal node Q. To balance the capacitance on both nodes of the SA in order
to minimize offset, an identical-valued capacitor MOSCap2 is placed on Qb but
connected to VDD such that it does not inject charge into node Qb. Figures 7.2
and 7.3 show *read* waveforms for "0" and "1," respectively; V(WL1) is the word
selection signal applied to the cells in the first-row of the memory array.

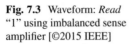

Fig. 7.3 Waveform: *Read* "1" using imbalanced sense amplifier [©2015 IEEE]

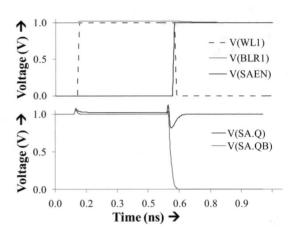

7.3   Charge-Injection-Based Single-Ended Imbalanced Sense Amplifier with Open-Bitline

This section presents single-ended sensing with imbalanced SA for an open-bitline architecture applicable to Dual-Port memory (DPSRAM) architecture [95]. The open-bitline architecture shown in Fig. 7.4 has SAs in the center with top and bottom bitlines connected to it, BLR_Top and BLR_Bot SA, respectively. *Read* can be performed on either top or bottom bitline as well as on both simultaneously taking advantage of the two SAs for every eight columns depending on the address to be read, see detail in Fig. 7.4.

In [95] this sensing method is applied with dual bitlines to implement dual *read* ports using the SA circuit in Fig. 7.5a. This SA is similar to that in Fig. 7.1 with half-bitlines BLR1_Top and BLR1_Bot being connected to each side of BLR1 SA; similarly BLR2_Top and BLR2_Bot are connected on each side of BLR2 SA. An imbalance of 100 mV is created using charge injection into SA nodes Q or Qb depending on the bitline to be sensed, BLR_Top or BLR_Bot, selected by activating either En_Top or En_Bot, respectively, see Fig. 7.5. In case of reading on BLR_Top or BLR_Bot, the other precharged bitline BLR_Bot or BLR_Top, respectively, works as reference voltage. Similar to Fig. 7.1, a bitline discharge of less than 100 mV is resolved by the SA as "1"; the bitline should discharge more than 150 mV for reliably reading "0" including the SA offset (50 mV).

This *read* scheme and architecture can be used for single-port memories as well using the SA circuit in Fig. 7.5a. In this case one of the BLs will be held at a reference voltage.

Figure 7.5b, shows the circuit used to drive the memory output during a *read* operation. Either the EN_Top or EN_Bot signal is activated with sense enable (SAEN) depending on whether the read cell is on BLR_Top or BLR_Bot to create an imbalance using charge injection into the SA (Fig. 7.5a) and to properly drive the read data output as shown in Fig. 7.5b. When the read cell is on BLR_Top, En_Top

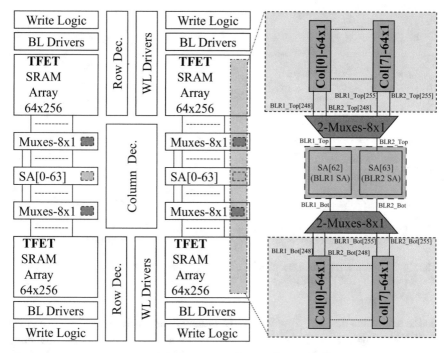

Fig. 7.4 DPSRAM memory: open-bitlines architecture [©2015 IEEE]

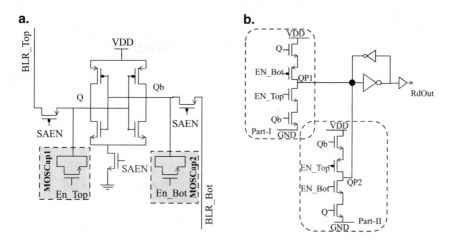

Fig. 7.5 Circuit diagram—(**a**) Sense amplifier, (**b**) Memory output driver [©2015 IEEE]

goes high with SAEN while En_Bot remains low. Signal QP1 in Part-I of the circuit shown in Fig. 7.5b drives the output RdOut while node QP2 in Part-II of the circuit is in tri-state. Similarly, for reading a cell on BLR_Bot, En_Top is low, En_Bot goes high with SAEN, QP1 is in tri-state and QP2 drives the output RdOut. Depending

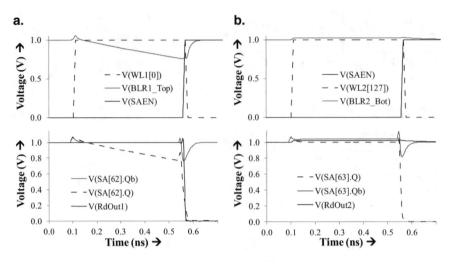

Fig. 7.6 Waveforms for simultaneous *read* on both ports (see Fig. 7.4, (**a**) Port-1 (read "0"), (**b**) Port-2 (read "1") [©2015 IEEE]

on the read value on BLR_Top, "0" or "1," either Qb or Q drive the output to a GND or VDD voltage, respectively. The latch is placed at the output to buffer the value read for the remaining period of the clock cycle. The feedback inverter of the latch is very weak as compared to the forward inverter to avoid having a tri-state inverter in the feedback path.

Figure 7.6 shows the waveforms for dual *read* on row[0] BLR1_Top at "0" and row[127] BLR2_Bot at "1" simultaneously of the same eight-column set. There are two SAs for each eight-column set, BLR1 SA and BLR2 SA; both BLR1 SA and BLR2 SA have as inputs BLR_Top and BLR_Bot. The waveforms in Fig. 7.6 show BLR1 SA reading data "0" from BLR1_Top which results in BLR1 SA node Q discharging; BLR2 SA is reading a "1" on BLR2_Bot keeping node Qb of the SA at the precharged value.

7.4   TFET *NDR* Skewed Inverter-Based Sensing Method

An ultra-compact skewed inverter-based sensing method, which makes use of the TFET *Negative Differential Resistance* (*NDR*) property is presented in this section [96]. The proposed approach simplifies the reading circuit and provides fast *read* speed by limiting the bitline discharge to < 200 mV even for an inverter-based *read* circuit.

Fig. 7.7 Proposed sense amplifier *Read* circuit [©2016 IEEE]

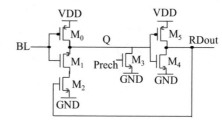

Fig. 7.8 DC characteristics with different sizing of M_0 with widths of 200 nm, 400 nm, 600 nm, and 800 nm [©2016 IEEE]

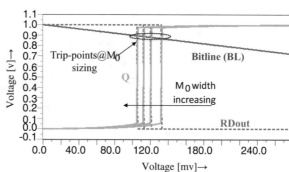

7.4.1   *Sensing Method*

The sense amplifier schematic is depicted in Fig. 7.7; it is based on *NDR* to skew the inverter threshold allowing limited voltage discharge on bitlines during *read* and enabling detection of bitline discharge of less than 200 mV by the skewed inverter. This eliminates the need of conventional voltage-latch-based SA and makes it easier to fit in the memory column pitch.

At the start of a *read* cycle NTFET M_1 is biased in reverse V_{DS}, $V_{DS} < 0$, while bitline BL and node Q are precharged to VDD (1 V) and GND, respectively, RDout at VDD, which keeps M_2 ON. During the *read* operation BL is discharging when connected to a "0," PTFET (M_0) attempts to charge node Q toward VDD while NTFET (M_1) resists due to the hump current (region-I of TFET characteristics), see Fig. 2.9, Chap. 2, until node Q is above 100 mV. Once the voltage on node Q is greater than 100 mV M_1 is in region II of the TFET characteristic, the hump current diminishes, and M_0 charges this node faster toward VDD. This results in a skewed trip point for Q with respect to BL: after $V_Q > 0.6$ V, M_1 starts having thermionic injection (region-III of the TFET characteristic) and will resist further charging of node Q. Therefore, the output inverter connected to Q providing RDout to drive the IO buffer/latch is designed with a trip point around $VDD/2$ switching OFF M_2 to stop the current through M_1 once RDout is resolved to "0", and allowing full charging of node Q to VDD. In order to meet the design specification of the memory, the trip point can be easily adjusted by changing the transistor sizings. Figure 7.8 shows the DC characteristics for different M_0 sizings, which result in different inverter (M_0 and M_1) trip points.

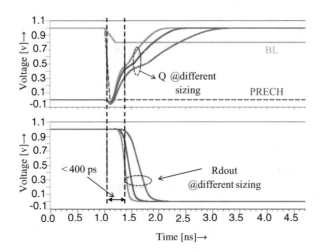

Fig. 7.9 Different trip points for sensing ($5fF$ load); M_0 sizing—400 nm blue; 600 nm pink; 800 nm orange [©2016 IEEE]

During *read* "0" operation with the wordline driven high and SA precharge OFF, BL is discharging, and at the same time the SA PTFET M_0 provides current charging node Q. Q starts charging because of $I_D(M_0) >> I_D(M_1)$ (M_0 ON current vs. M_1 hump current). As explained before, Q charges faster once $V_Q > 100$ mV. However, the full TFET circuit is still slower in fully charging and discharging Q and RDout, respectively, due to its lower current drivability in comparison to MOSFETs. The speed can be improved by using MOSFETs for M_0, M_4, and M_5. As shown in Fig. 7.9, a bitline discharge of 200 mV at 1 V supply is sufficient for reading "0" by the hybrid TFET-CMOS circuit. This is difficult to achieve for standard CMOS or TFET inverter, i.e., without using *NDR* for TFET, even with sizing optimized to increase the skew. For reading "1," BL is maintained precharged at VDD, thus M_0 is OFF and NTFET M_1 preserves "0" on Q because of the hump current. RDout remains at VDD and thus M_2 remains ON. The *read* delay for different supply voltages is depicted in Fig. 7.10 demonstrating the good *read* scalability of the circuit.

7.4.2 Summary

A TFET *NDR*-based skewed inverter *read* scheme is promising to detect bitline discharge from 150 mV to 300 mV with an optimized area independent of the value of VDD. The proposed TFET-CMOS circuit provides a *read* delay from 393 ps to 904 ps for a supply voltage ranging from 1 V to 0.6 V, respectively, with a load of $5fF$ on RDout. This circuit proves to be an area-efficient single-ended *read* circuit

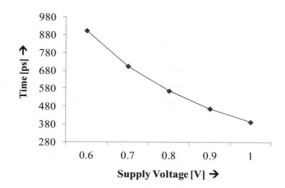

Fig. 7.10 *Read* delay vs. supply voltage for TFET-CMOS SA with a $5fF$ load [©2016 IEEE]

providing an operation speed similar to the one of a conventional differential latch-based sense amplifier.

7.5 Adaptive Read Technique

This section presents the design and operation of an adaptive *read* technique for memories supporting Advance Voltage Scaling (AVS) with in-situ detection of end-of-correct *read* operation. This technique is applied to the design of an adaptive sense amplifier and it is used in synchronous and asynchronous SRAM architectures using CMOS, RRAM, TFETs, or any other memory technology.

In our work we target hybrid TFET-CMOS memories combining the best features of TFET with those of CMOS. TFETs are best suited for implementing the storage cells due to their reduced leakage while MOSFETs are best for the periphery circuits due to their higher ON current and speed. The circuits described in this section are designed in CMOS as the focus is on sensing.

7.5.1 Adaptive Read Technique Operation

The adaptive *read* technique can detect the voltage, at which the bitline discharge is sufficient for a reliable *read*. This is achieved by monitoring the bitline discharge before the wordline is deselected. Figure 7.11 shows the SA schematic for this approach; BL and BLB (precharged at VDD) are connected to the gates of pass transistors PG1 and PG2. The operation of the SA is as follows. During precharge nodes SP1 and SP2 are discharged to GND by the Precharge circuit in Fig. 7.11c; during *read*, SAEN and Prech are low. The differential voltage (VdiffSP) generated between SP1 and SP2 is the amplified version of the differential voltage (Vdiff) between BL and BLB. Initially Q1 and Q2 are high; with increasing Vdiff due to increasing voltage difference between bitlines during *read*, either Q1 or Q2 starts

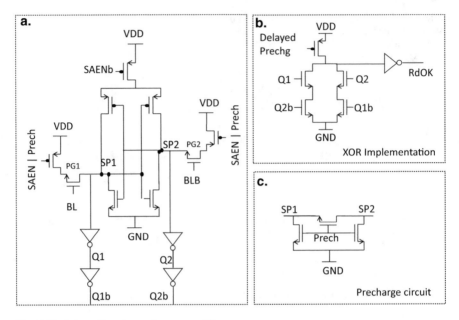

Fig. 7.11 Schematic: proposed sense amplifier

discharging toward zero (depending on whether BL or BLB is discharging), and the other node tries to remain near VDD. The four output signals of the SA, Q1, Q2, Q1b, and Q2b are inputs to the XOR gate represented in Fig. 7.11b. A sufficiently large Vdiff triggers the SA into "1-0" or "0-1" state, which in turn through the signals applied to the XOR generates the RdOK high only when Q1 and Q2 are resolved, i.e., the voltage on one is close to VDD and the other close to GND. If RdOK does not go high before the wordline is deselected, it signifies that the WL pulse-width is not long enough to read and the read value is not reliable. In other words, Vdiff between BL and BLB is not sufficient to perform a correct *read*. As shown in Fig. 7.12, this scheme can be used in designing adaptive memories to notify the end of a correct *read* operation; the RdOK signal can be used to switch WL OFF and to switch ON the latch behavior of the SA to preserve the *read* value. This is achieved by performing the logic functions shown in Fig. 7.12, "SAEN OR RdOK OR Prech," and, "SAENb AND RdOKb." Using this scheme various memory architectures can be defined either with control logic implemented inside or outside the memory.

7.5.2 Detailed Description of Functionality

Below is the description of the *read* operations using the SA shown in Fig. 7.11. At the start of the *read*, i.e., selected WL goes high and Prech going low after a delay,

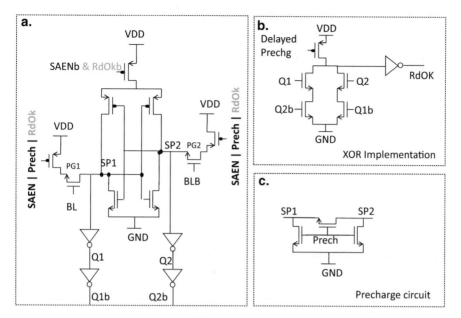

Fig. 7.12   Schematic: self-adaptive sense amplifier

SP1 and SP2 charge to an intermediate value with VdiffSP of 0 V. VdiffSP between SP1 and SP2 starts increasing as the voltage on SP1 starts decreasing once BL starts discharging; for example, for Vdiff of 100 mV, VdiffSP can be larger than 300 mV through positive feedback in SA core. Q1 either starts charging back to VDD, if Q1 gets a glitch when Prech is switched OFF, or retains a value near VDD; Q2 starts discharging to GND. During the start of the operation RdOK is "0." With Q1 and Q2 resolving to VDD and GND, respectively, RdOK starts charging to VDD. A sufficient Vdiff is defined as the differential voltage between BL and BLB, which generates $VdiffSP > Voffset$ of the SA. For example, the SA can be designed with $VdiffSP \geq VDD/2$, and transistor sizing is done such that Q1 and Q2 are close to GND or VDD for $VdiffSP \geq VDD/2$. As soon as Q1 and Q2 are resolved, RdOK goes high notifying that the *read* operation is finished and the read value is reliable. A low value on RdOK at the end of the *read*, i.e., WL low, SAEN high, and SAENb low, signifies that the *read* is unreliable and the differential voltage Vdiff was not sufficient for a proper *read* operation.

At the level of an $n * m$ memory array the adaptive *read* technique can be implemented in different ways; we describe three architectures in the following subsections.

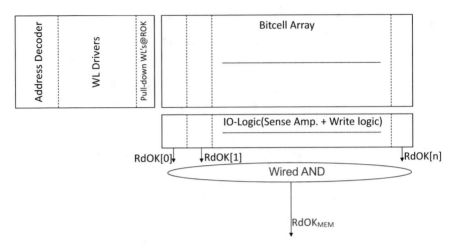

Fig. 7.13 Memory architecture with $RdOK_{MEM}$ generation

7.5.2.1   Architecture-1: RdOK to Notify Reliability of *Read* Operation

In the first proposed architecture the RdOK signals from all the read columns are wire ANDed and a $RdOK_{MEM}$ signal is generated as shown in Fig. 7.13. This signal is provided as an output from the memory. This can be utilized by the system control logic to adapt the system in case of an unreliable *read* operation. For example, if RdOK is low at the end of the *read* operation, the *read* is repeated with increased supply voltage. The control can be implemented in hardware or it can be implemented in software running on the system processor. The RdOK signal generation from a column correctly read is shown in Fig. 7.14; the waveform for an unreliable *read* is shown in Fig. 7.15.

7.5.2.2   Architecture-2: SRAM with Self-Tuning Using RdOK

In this architecture the RdOK signals from all the read columns are used by control logic inside the memory, see Fig. 7.16. The control logic can adapt the memory in case of an unreliable *read* operation; for example, if any of the RdOK's is low at the end of the *read* operation, the *read* is repeated with increased supply voltage, lengthened WL pulse-width, or with any kind of *read-assist (RA)* technique. The $RdOK_{MEM}$ is also generated to notify the system using the memory that the last *read* operation was unreliable and needs to be repeated. This control can be implemented in BIST, such that during normal operation if any unreliable *read* is detected, the memory BIST starts and tunes the memory.

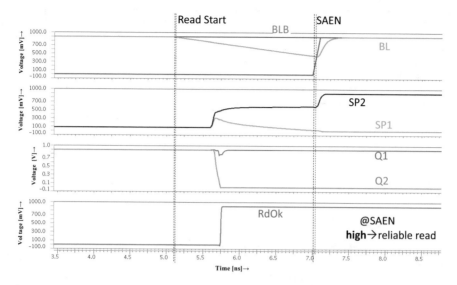

Fig. 7.14 Waveforms: reliable *Read* operation

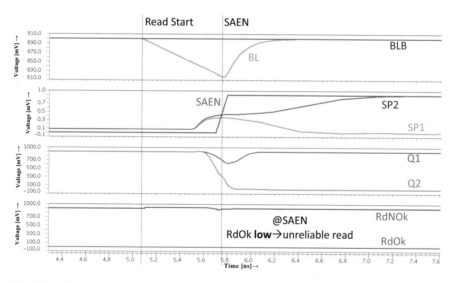

Fig. 7.15 Waveforms: unreliable *Read* operation

7.5.2.3 Architecture-3: Self-Adapting SRAM Using RdOK

In this architecture control logic inside the memory processes the RdOK signal from all the read columns. As shown in Fig. 7.17, the control logic is designed to use RdOK signals to decide the end of a correct *read* operation and to pull down WL as soon as RdOK signals are generated from all read columns. Waveforms for adaptive

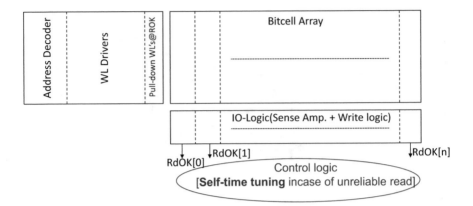

Fig. 7.16 Memory architecture with self-tuning

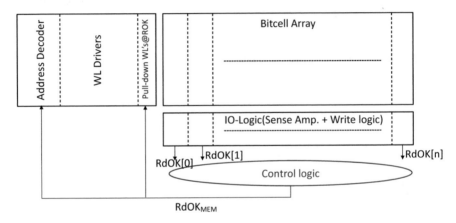

Fig. 7.17 Self-adapt memory architecture

read are shown in Fig. 7.18. This architecture based on the SA shown in Fig. 7.12, is used when designing in-situ optimized adaptive WL pulse-width. The control logic can also be designed to consume two clock cycles in case of insufficient bitline discharge at the end of the first cycle. For example, if any of the RdOK's is low at the end of the clock period, WL is not pulled down and bitline discharge continues further until $RdOK_{MEM}$ is generated correctly. A single-bit output signal is provided to the system to notify that the memory needs more than one cycle for the ongoing *read* operation. If $RdOK_{MEM}$ is still not generated at the end of the second clock cycle, the operation is ended with the signal notifying unreliable *read* and the higher-level system outside the memory has to take control for correction and has to repeat the *read* by increasing the supply voltage. This can be used in systems implementing adaptive voltage scaling, especially for embedded memories in processors.

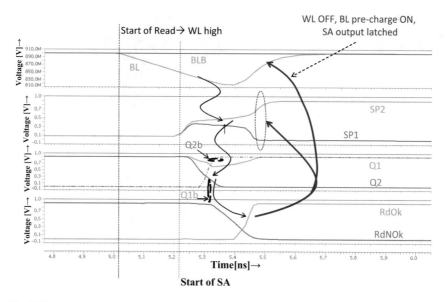

Fig. 7.18 Waveforms: adaptive *Read*

7.5.3 Alternative Design Options

7.5.3.1 Asynchronous Memory Design

In asynchronous memories SA timing is internally generated using for example a dummy read column to generate the self-timed SAEN signal. The presented SA and adaptive memory architecture are applicable to asynchronous memories. Since the SA can decide the end of a correct *read* operation, the *read* can be performed without the clock signal to generate SAEN; $RdOK_{MEM}$ can be used to generate *end the read operation* timing as shown in Fig. 7.17 and to send the data to the output.

7.5.3.2 Single-Ended Sensing

The described sense amplifier and adaptive memory architecture concept are also applicable for single-ended sensing with bitline precharge at $VDD/2$ or VDD.

- In the case of bitline precharged at $VDD/2$, an SA architecture similar to that in Fig. 7.12 is used; the bitline is split into Bitline Top (BLT) and Bitline Bot (BLB) in a memory array architecture like the one in Fig. 7.4; the bitcell is read using either BLT or BLB with the other bitline precharged at $VDD/2$. Depending on whether the read bitline is charged or discharged during the *read* operation, the read value can be either "1" or "0," respectively.

Fig. 7.19 Schematic:
adaptive single-ended sense
amplifier

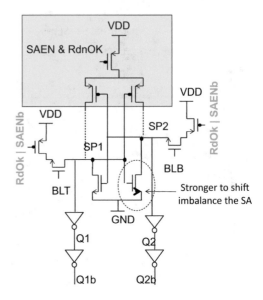

- In the case of bitline precharged at VDD, the self-adaptive single-ended SA shown in Fig. 7.19 is used; RdnOk is the inverse of RdOk signal; the SA is imbalanced in order to perform a single-ended *read* on BLT while using BLB precharged at VDD as reference voltage; the SA is imbalanced to favor BLT, i.e., if BLT does not discharge during *read*, the SA senses a "1." Imbalance in SA can be introduced in various ways; for example, it can be introduced by sizing asymmetrically the NMOS differential pair in the SA or by changing the precharge voltage of BLB.

7.5.4 Summary and Applications

The introduced SA architecture for adaptive *read* is applicable for all the use-cases where reliability or voltage scaling is required. Important applications include embedded systems such as cache memories of processors and memories used in SoCs designed for the Internet of Things, machine learning, etc. The focus of this chapter was the design of TFET-CMOS hybrid memories with sensing circuits implemented in CMOS in order to optimize device dimensions for the same speed as TFETs, and benefit from accurate transistor variability models available for CMOS but unavailable for TFETs. As a proof of concept an SRAM macro with adaptive sensing technique was designed in 28 nm FDSOI CMOS due to limitations imposed to circuit complexity by present TFET processes. The layout of the 8 Kb test-macro implemented in a 28 nm FDSOI-CMOS process is shown in Fig. 7.20.

Fig. 7.20 Adaptive memory
test-chip photo and layout
details

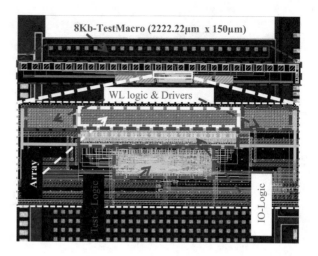

Chapter 8
Perspective

8.1 TFET-CMOS Hybrid Cores

The previous chapters have demonstrated the extraordinary potential of implementing processors and digital ICs using TFETs in today's world of mobile devices operating on meager power budgets. The promise of TFET devices does not stem only from their own characteristics but from their compatibility and integration with CMOS devices on the same chip. The unique characteristics of the TFET led us to imagine a few extremely power efficient and compact new circuit topologies described in this book.

In order to envisage TFET or hybrid applications the first consideration is the type of the TFET device to be used. The majority of recent investigations focus on the use of heterojunction TFETs (HTFET) since the intent is to match or overtake the performance of currently used MOS devices while decreasing the leakage. The reason behind this approach is that TFETs will replace MOSFETs. However, as mentioned in Chap. 2 and further explored in the following chapters HTFET devices suffer from a number of problems. First is the difficulty in fabrication, which makes it a major challenge for foundries to manufacture and classify the device as viable for large-scale integration and production. Moreover, as detailed in Chap. 2, due to the introduction of additional process variation sources, the problem is becoming even worse. Second, even if the HTFET device could be used it has its limitations in circuit applications due to the turn-on of the pin diode at very low reverse bias, much lower than in the case of Si-based devices. This poses major challenges to any design where a device is supposed to be symmetric and conduct current in both directions, like in the vast majority of CMOS memory circuits, as the circuit techniques applicable to Si-TFETs to mitigate diode current at reverse bias, are very limited.

Silicon TFETs, on the other hand, offer lower drive current but also provide reduced leakage and are less constrained in reverse-bias operation as detailed in Chap. 2. Compared to HTFETs the fabrication in silicon is straightforward as it is

© Springer Nature Switzerland AG 2021

N. Gupta et al., *TFET Integrated Circuits*,

https://doi.org/10.1007/978-3-030-55119-3_8

compatible with current CMOS processes and can have reasonable production yield. Moreover, since these devices are purely silicon based, on-chip cointegration with standard CMOS devices is possible.

In recent years the market of battery-powered devices has been experiencing year-to-year exponential growth, a trend that is expected to continue. Within this market there are many devices, different kinds of sensors, for example, that are event-triggered with long standby cycles and have low operating frequency requirements while active. The standard approach to address this operating mode is to introduce sleep intervals for memories where the periphery is power gated and bitcell arrays are kept at scaled down supply voltages. The gains offered by this technique are, however, limited, moreover they introduce additional design and timing complexity and the requirement to generate and maintain additional supply voltage levels. Another solution consists in the use of non-volatile devices, such as RRAM, PCM, MRAM, etc., which maintain data even if the memory is completely powered down. The problem with this approach is that non-volatile devices suffer from a range of problems like the limited production yield, low endurance, high variability (also cycle-to-cycle), high programming currents, low performance, etc., placing them in the memory hierarchy somewhere between DRAM and Flash, rather than replacing SRAMs.

In battery-powered applications TFETs are an obvious match since the extremely low leakage with reduced drive current fits the requirements perfectly. Moreover, sensor-node designs are highly area constrained as the cost is one of the key concerns. It was demonstrated that TFETs can be competitive or superior even in that regard either by the use of a standard 6T-SRAM-bitcell based design or with the more advanced designs such as the 3T SRAM (see Chap. 3) or the uDRAM-based 2T1C SRAM (see Chap. 4). With the use of the TFET Flip-flop (see Chap. 5) and having designed a set of TFET standard cells a full Si-TFET SoC can be built. Given the properties of the device, this can be a perfect solution for systems where leakage and area are targeted with lower performance (in the range of tens of MHz) at nominal voltage, as is the case for many sensor-node applications.

Medium-to-high performance, low leakage, and low energy pure Si-TFET cores are, however, challenging. As detailed in Chap. 2, TFETs exhibit rather poor PDP and suffer in performance from device stacking due to high $I_D(V_{DS})$ dependence, making logic gates with a few transistors stacked between power and ground very slow. It was demonstrated though, that memories and flip-flops can be designed in a way, which provides high frequency and low leakage, even though the standard-cell logic might be a bottleneck. One of the solutions to the problem is to design hybrid TFET-CMOS cores. In [4] a hybrid TFET-CMOS multi-core processor is presented using CMOS cores for higher voltage and TFET cores for lower voltage of operation. This is done to optimize energy efficiencies based on the computation loads. Presented results show a 50% energy benefit and 25% energy-delay product (EDP) benefit with single-threaded applications and up to 55% EDP benefit with multi-threaded applications. Various benchmarks were run for evaluating the performance. In [5], EDP-aware and barrier-aware DVFS is used to maximize the use of TFET cores for energy efficiency. The presented results show up to 30% leakage and 17% dynamic power savings with a performance degradation of only 1%. However, in the proposed system, either CMOS cores or TFET cores are unused at any given point of time resulting in unused silicon footprint and area, which leads to increased cost of the system. Assuming the same size of the cores, 50% silicon area will be unused at any given point of time. In addition, due to higher than CMOS C_{GD}, the dynamic power consumption of TFET cores will be higher than CMOS for the same voltage of operation, which reduces the gain in dynamic power consumption significantly.

A major improvement to existing art would be a tighter cointegration of TFET and CMOS components. It was demonstrated in this book that exploiting the *NDR* property of Si-TFETs, low-leakage, high-performance memories, and flip-flops are achievable. In a hybrid TFET-CMOS core it would be therefore beneficial to share TFET memories, SRAM, and uDRAM if necessary, between TFET and CMOS cores and reuse TFET flip-flops in both.

Another dimension of a hybrid system involves an even tighter cointegration of TFET and CMOS within the system. Provided the TFET and CMOS standard cells can be cointegrated in the same design with no or low area overhead it would be possible to design full TFET-CMOS hybrid cores. Even if only a reduced set of TFET standard cells with full timing specification is provided to the synthesis and place-and-route EDA tools, the software itself will be able to decide which type of cell to use in different parts of the logic block. This could result, for example, in the clock tree being implemented using fast and low PDP CMOS gates while large parts of the remaining blocks being designed with TFETs in order to keep the static power low. Such a system together with TFET memories could be a powerful solution for many applications since it should provide sufficiently high operating frequency coupled with extremely low standby power due to low-leakage memories and the fact that the logic core can be fully power-gated if necessary. Flip-flops can be kept on the secondary power grid and kept powered up even during standby in the event where their states have to be maintained; this would come at a certain area cost. Even in this situation, however, their contribution to the leakage power remains negligible (see Chap. 5) since these flip-flops are TFET based. It should also be noted that this kind of hybrid system could operate at a nominal voltage since there is no need for different voltage domains for dual-rail memory management, as is typically the case in many modern applications. This simplifies the power routing and reduces area footprint and power consumption relative to the use of DC–DC converters.

There are multiple challenges that would have to be addressed delivering the above scenario. First is the design and full characterization of all the standard cells and memory IPs required for the implementation of such a system. Next would be the cointegration with different types of existing devices and benchmarking the performance for various types of applications and algorithms. Also, should the results prove significantly better than the existing art one of the foundries would need to invest in the technology. Last, but not least, our technologists will need to deliver on the promise of a 30 mV/decade integrated TFET. Given the exponential growth of the battery-powered device market, the investment might be well worth the effort.

References

1. Costin Anghel, Prathyusha Chilagani, Amara Amara, and Andrei Vladimirescu. Tunnel field effect transistor with increased ON current, low-k spacer and high-k dielectric. *Applied Physics Letters*, 96(12):122104, 2010.
2. Massimo Alioto. Ultra-low power design approaches for IoT. In *Hot Chips Symposium*, pages 1–57, 2014.
3. ITRS Organization. International Technology Roadmap for Semiconductors Roadmap.
4. Vinay Saripalli, Asit Mishra, Suman Datta, and Vijaykrishnan Narayanan. An energy-efficient heterogeneous CMP based on hybrid TFET-CMOS cores. In *Design Automation Conference (DAC), 2011 48th ACM/EDAC/IEEE*, pages 729–734. IEEE, 2011.
5. Karthik Swaminathan, Emre Kultursay, Vinay Saripalli, Vijaykrishnan Narayanan, Mahmut Kandemir, and Suman Datta. Improving energy efficiency of multi-threaded applications using heterogeneous CMOS-TFET multicores. In *Proceedings of the 17th IEEE/ACM international symposium on Low-power electronics and design*, pages 247–252. IEEE Press, 2011.
6. Supreet Jeloka, Naveen Akesh, Dennis Sylvester, and David Blaauw. A configurable TCAM/BCAM/SRAM using 28nm push-rule 6T bit cell. In *VLSI Circuits (VLSI Circuits), 2015 Symposium on*, pages C272–C273. IEEE, 2015.
7. Stefan Rusu, Harry Muljono, David Ayers, Simon Tam, Wei Chen, Aaron Martin, Shenggao Li, Sujal Vora, Raj Varada, and Eddie Wang. A 22 nm 15-Core Enterprise Xeon® Processor Family. *IEEE Journal of Solid-State Circuits*, 50(1):35–48, 2015.
8. Toshikazu Fukuda, Koji Kohara, Toshiaki Dozaka, Yasuhisa Takeyama, Tsuyoshi Midorikawa, Kenji Hashimoto, Ichiro Wakiyama, Shinji Miyano, and Takehiko Hojo. A 7ns-access-time 25μW/MHz 128kb SRAM for low-power fast wake-up MCU in 65nm CMOS with 27fA/b retention current. In *Solid-State Circuits Conference Digest of Technical Papers (ISSCC), 2014 IEEE International*, pages 236–237. IEEE, 2014.
9. James Myers, Anand Savanth, David Howard, Rohan Gaddh, Pranay Prabhat, and David Flynn. 8.1 An 80nW retention 11.7 pJ/cycle active subthreshold ARM Cortex-M0+ subsystem in 65nm CMOS for WSN applications. In *Solid-State Circuits Conference-(ISSCC), 2015 IEEE International*, pages 1–3. IEEE, 2015.
10. Mahmood Khayatzadeh, Xiaoyang Zhang, Jun Tan, Wen-Sin Liew, and Yong Lian. A 0.7-v 17.4-/spl mu/w 3-lead wireless ecg soc. *IEEE transactions on biomedical circuits and systems*, 7(5):583–592, 2013.
11. Scott Hanson, Mingoo Seok, Yu-Shiang Lin, ZhiYoong Foo, Daeyeon Kim, Yoonmyung Lee, Nurrachman Liu, Dennis Sylvester, and David Blaauw. A low-voltage processor for sensing applications with picowatt standby mode. *IEEE Journal of Solid-State Circuits*, 44(4):1145–1155, 2009.

© Springer Nature Switzerland AG 2021
N. Gupta et al., *TFET Integrated Circuits*,
https://doi.org/10.1007/978-3-030-55119-3

12. J Appenzeller, Y-M Lin, J Knoch, and Ph Avouris. Band-to-band tunneling in carbon nanotube field-effect transistors. *Physical review letters*, 93(19):196805, 2004.

13. Woo Young Choi, Byung-Gook Park, Jong Duk Lee, and Tsu-Jae King Liu. Tunneling field-effect transistors (TFETs) with subthreshold swing (SS) less than 60 mV/dec. *IEEE Electron Device Letters*, 28(8):743–745, 2007.

14. Costin Anghel, Anju Gupta, Amara Amara, Andrei Vladimirescu, et al. 30-nm tunnel FET with improved performance and reduced ambipolar current. *IEEE Transactions on Electron Devices*, 58(6):1649–1654, 2011.

15. A Villalon, C Le Royer, P Nguyen, S Barraud, F Glowacki, A Revelant, L Selmi, S Cristoloveanu, L Tosti, C Vizioz, et al. First demonstration of strained SiGe nanowires TFETs with ION beyond $700\mu A/\mu m$. In *VLSI Technology (VLSI-Technology): Digest of Technical Papers, 2014 Symposium on*, pages 1–2. IEEE, 2014.

16. Ramanathan Gandhi, Zhixian Chen, Navab Singh, Kaustav Banerjee, and Sungjoo Lee. Vertical Si-Nanowire n-Type Tunneling FETs With Low Subthreshold Swing (leq 50mV/decade) at Room Temperature. *IEEE Electron Device Letters*, 32(4):437–439, 2011.

17. David Esseni, Manuel Guglielmini, Bernard Kapidani, Tommaso Rollo, and Massimo Alioto. Tunnel FETs for ultralow voltage digital VLSI circuits: Part I Device–circuit interaction and evaluation at device level. *IEEE Transactions on Very Large Scale Integration (VLSI) Systems*, 22(12):2488–2498, 2014.

18. R Ranica, N Planes, O Weber, O Thomas, S Haendler, D Noblet, D Croain, C Gardin, and F Arnaud. FDSOI process/design full solutions for ultra low leakage, high speed and low voltage SRAMs. In *VLSI Technology (VLSIT), 2013 Symposium on*, pages T210–T211. IEEE, 2013.

19. Hraziia, Andrei Vladimirescu, Amara Amara, and Costin Anghel. An analysis on the ambipolar current in Si double-gate tunnel FETs. *Solid-State Electronics*, 70:67–72, 2012.

20. Adrian M Ionescu and Heike Riel. Tunnel field-effect transistors as energy-efficient electronic switches. *nature*, 479(7373):329, 2011.

21. YM Niquet, C Delerue, G Allan, and M Lannoo. Method for tight-binding parametrization: Application to silicon nanostructures. *Physical Review B*, 62(8):5109, 2000.

22. Kathy Boucart and Adrian Mihai Ionescu. Double-Gate Tunnel FET With High-K Gate Dielectric. *IEEE Transactions on Electron Devices*, 54(7):1725–1733, 2007.

23. C Sandow, J Knoch, C Urban, Q-T Zhao, and S Mantl. Impact of electrostatics and doping concentration on the performance of silicon tunnel field-effect transistors. *Solid-State Electronics*, 53(10):1126–1129, 2009.

24. Navneet Gupta, Adam Makosiej, Costin Anghel, Amara Amara, and Andrei Vladimirescu. Cmos sensor nodes with sub-picowatt TFET memory. *IEEE Sensors Journal*, 16(23):8255–8262, 2016.

25. ASU. Predictive Technology Model, 1999.

26. Vinay Saripalli, Suman Datta, Vijaykrishnan Narayanan, and Jaydeep P Kulkarni. Variation-tolerant ultra low-power heterojunction tunnel FET SRAM design. In *Proceedings of the 2011 IEEE/ACM International Symposium on Nanoscale Architectures*, pages 45–52. IEEE Computer Society, 2011.

27. Sneh Saurabh and M Jagadesh Kumar. Estimation and compensation of process-induced variations in nanoscale tunnel field-effect transistors for improved reliability. *IEEE Transactions on Device and Materials Reliability*, 10(3):390–395, 2010.

28. Marcel JM Pelgrom and Aad CJ Duinmaijer. Matching properties of MOS transistors. In *ESSCIRC'88: Fourteenth European Solid-State Circuits Conference*, pages 327–330. IEEE, 1988.

29. Saurabh Mookerjea, Ramakrishnan Krishnan, Suman Datta, and Vijaykrishnan Narayanan. On enhanced Miller capacitance effect in interband tunnel transistors. *IEEE Electron Device Letters*, 30(10):1102–1104, 2009.

30. Jawar Singh, Krishnan Ramakrishnan, S Mookerjea, Suman Datta, Narayanan Vijaykrishnan, and D Pradhan. A novel si Tunnel-FET-based SRAM design for ultra low-power 0.3V VDD applications. In *Proceedings of the 2010 Asia and South Pacific Design Automation Conference*, pages 181–186. IEEE Press, 2010.

31. Daeyeon Kim, Yoonmyung Lee, Jin Cai, Isaac Lauer, Leland Chang, Steven J Koester, Dennis Sylvester, and David Blaauw. Low power circuit design based on heterojunction tunneling transistors (HETTs). In *Proceedings of the 2009 ACM/IEEE international symposium on Low power electronics and design*, pages 219–224. ACM, 2009.

32. Adam Makosiej, Rutwick Kumar Kashyap, Andrei Vladimirescu, Amara Amara, and Costin Anghel. A 32nm tunnel FET SRAM for ultra low leakage. In *Circuits and Systems (ISCAS), 2012 IEEE International Symposium on*, pages 2517–2520. IEEE, 2012.

33. Alan C Seabaugh and Qin Zhang. Low-voltage tunnel transistors for beyond CMOS logic. *Proceedings of the IEEE*, 98(12):2095–2110, 2010.

34. Xuebei Yang and Kartik Mohanram. Robust 6T Si tunneling transistor SRAM design. In *Design, Automation & Test in Europe Conference & Exhibition (DATE), 2011*, pages 1–6. IEEE, 2011.

35. Jan M Rabaey, Anantha P Chandrakasan, and Borivoje Nikolic. *Digital Integrated Circuits, A Design Perspective*. Prentice hall Englewood Cliffs, Second edition, 2002.

36. Leland Chang, David M Fried, Jack Hergenrother, Jeffrey W Sleight, Robert H Dennard, Robert K Montoye, Lidija Sekaric, Sharee J McNab, Anna W Topol, Charlotte D Adams, et al. Stable SRAM cell design for the 32 nm node and beyond. In *VLSI Technology, 2005. Digest of Technical Papers. 2005 Symposium on*, pages 128–129. IEEE, 2005.

37. O Thomas, C Anghel, and Adam Makosiej. Cellule memoire a transistors de lecture de type TFET et MOSFET, 2014. European Patent Application EP3010022A1.

38. V Saripalli et al. Generic TFET based 4T memory devices. *US Patent-Issued, No. US8638591*, 2014.

39. Leland Chang, Yutaka Nakamura, Robert K Montoye, Jun Sawada, Andrew K Martin, Kiyofumi Kinoshita, Fadi H Gebara, Kanak B Agarwal, Dhruva J Acharyya, Wilfried Haensch, et al. A 5.3 GHz 8T-SRAM with operation down to 0.41 V in 65nm CMOS. In *VLSI Circuits, 2007 IEEE Symposium on*, pages 252–253. IEEE, 2007.

40. Yasuhiro Morita, Hidehiro Fujiwara, Hiroki Noguchi, Yusuke Iguchi, Koji Nii, Hiroshi Kawaguchi, and Masahiko Yoshimoto. An area-conscious low-voltage-oriented 8T-SRAM design under DVS environment. In *VLSI Circuits, 2007 IEEE Symposium on*, pages 256–257. IEEE, 2007.

41. S Yoshimoto, M Terada, S Okumura, T Suzuki, S Miyano, H Kawaguchi, and M Yoshimoto. A 40-nm 0.5-V 20.1-μW/MHz 8T SRAM with low-energy disturb mitigation scheme. In *VLSI Circuits (VLSIC), 2011 Symposium on*, pages 72–73. IEEE, 2011.

42. Jose Maiz, Scott Hareland, Kevin Zhang, and Patrick Armstrong. Characterization of multi-bit soft error events in advanced SRAMs. In *Electron Devices Meeting, 2003. IEDM'03 Technical Digest. IEEE International*, pages 21–4. IEEE, 2003.

43. Peter Hazucha, T Karnik, J Maiz, S Walstra, B Bloechel, J Tschanz, G Dermer, S Hareland, P Armstrong, and S Borkar. Neutron soft error rate measurements in a 90-nm CMOS process and scaling trends in SRAM from 0.25-/spl mu/m to 90-nm generation. In *Electron Devices Meeting, 2003. IEDM'03 Technical Digest. IEEE International*, pages 21–5. IEEE, 2003.

44. Huichu Liu, Matthew Cotter, Suman Datta, and Vijay Narayanan. Technology assessment of Si and III-V FinFETs and III-V tunnel FETs from soft error rate perspective. In *Electron Devices Meeting (IEDM), 2012 IEEE International*, pages 25–5. IEEE, 2012.

45. Huichu Liu, Matthew Cotter, Suman Datta, and Vijaykrishnan Narayanan. Soft-error performance evaluation on emerging low power devices. *IEEE Transactions on Device and Materials Reliability*, 14(2):732–741, 2014.

46. Hiroki Noguchi, Shunsuke Okumura, Yusuke Iguchi, Hidehiro Fujiwara, Yasuhiro Morita, Koji Nii, Hiroshi Kawaguchi, and Masahiko Yoshimoto. Which is the Best Dual-Port SRAM in 45-nm Process Technology? 8T, 10T single end, and 10T differential. In *Integrated Circuit Design and Technology and Tutorial, 2008. ICICDT 2008. IEEE International Conference on*, pages 55–58. IEEE, 2008.

47. Jui-Jen Wu, Meng-Fan Chang, Shau-Wei Lu, Robert Lo, and Quincy Li. A 45-nm dual-port SRAM utilizing write-assist cells against simultaneous access disturbances. *IEEE Transactions on Circuits and Systems II: Express Briefs*, 59(11):790–794, 2012.

48. Koji Nii, Yasumasa Tsukamoto, Tomoaki Yoshizawa, S Imaolka, and Hiroshi Makino. A 90nm dual-port SRAM with 2.04/spl mu/m/sup 2/8T-thin cell using dynamically-controlled column bias scheme. In *Solid-State Circuits Conference, 2004. Digest of Technical Papers. ISSCC. 2004 IEEE International*, pages 508–543. IEEE, 2004.

49. Koji Nii, M Yabuuchi, Y Tsukamoto, S Ohbayashi, Y Oda, K Usui, T Kawamura, N Tsuboi, T Iwasaki, K Hashimoto, et al. A 45-nm single-port and dual-port SRAM family with robust read/write stabilizing circuitry under DVFS environment. In *VLSI Circuits, 2008 IEEE Symposium on*, pages 212–213. IEEE, 2008.

50. Navneet Gupta, Adam Makosiej, Oliver Thomas, Amara Amara, Andrei Vladimirescu, and Costin Anghel. Ultra-low leakage sub-32nm TFET/CMOS hybrid 32kb pseudo Dual Port scratchpad with GHz speed for embedded applications. In *Circuits and Systems (ISCAS), 2015 IEEE International Symposium on*, pages 597–600. IEEE, 2015.

51. Ming-Hsien Tu, Jihi-Yu Lin, Ming-Chien Tsai, Shyh-Jye Jou, and Ching-Te Chuang. Single-ended subthreshold SRAM with asymmetrical write/read-assist. *IEEE Transactions on Circuits and Systems I: Regular Papers*, 57(12):3039–3047, 2010.

52. Richard F Hobson. A new single-ended SRAM cell with write-assist. *IEEE Transactions on Very Large Scale Integration (VLSI) Systems*, 15(2):173–181, 2007.

53. Jihi-Yu Lin, Ming-Hsien Tu, Ming-Chien Tsai, Shyh-Jye Jou, and Ching-Te Chuang. Asymmetrical Write-assist for single-ended SRAM operation. In *SOC Conference, 2009. SOCC 2009. IEEE International*, pages 101–104. IEEE, 2009.

54. Navneet Gupta, Adam Makosiej, Andrei Vladimirescu, Amara Amara, and Costin Anghel. 3T-TFET bitcell based TFET-CMOS hybrid SRAM design for ultra-low power applications. In *Design, Automation & Test in Europe Conference & Exhibition (DATE), 2016*, pages 361–366. IEEE, 2016.

55. K Hijioka, N Inoue, I Kume, J Kawahara, N Furutake, H Shirai, T Itoh, T Ogura, K Kazama, Y Yamamoto, et al. A novel cylinder-type MIM capacitor in porous low-k film (CAPL) for embedded DRAM with advanced CMOS logics. In *Electron Devices Meeting (IEDM), 2010 IEEE International*, pages 33–3. IEEE, 2010.

56. John Barth, Don Plass, Erik Nelson, Charlie Hwang, Gregory Fredeman, Michael Sperling, Abraham Mathews, Toshiaki Kirihata, William R Reohr, Kavita Nair, et al. A 45 nm SOI embedded DRAM macro for the POWER^{TM} processor 32 MByte on-chip L3 cache. *IEEE Journal of Solid-State Circuits*, 46(1):64–75, 2011.

57. Sergey Romanovsky, Atul Katoch, Arun Achyuthan, Cormac O'Connell, Sreedhar Natarajan, Chris Huang, Chuan-Yu Wu, Min-Jer Wang, CJ Wang, Paul Chen, et al. A 500MHz random-access embedded 1Mb DRAM macro in bulk CMOS. In *Solid-State Circuits Conference, 2008. ISSCC 2008. Digest of Technical Papers. IEEE International*, pages 270–612. IEEE, 2008.

58. Fatih Hamzaoglu, Umut Arslan, Nabhendra Bisnik, Swaroop Ghosh, Manoj B Lal, Nick Lindert, Mesut Meterelliyoz, Randy B Osborne, Joodong Park, Shigeki Tomishima, et al. A 1 Gb 2 GHz 128 Gb/s bandwidth embedded DRAM in 22 nm tri-gate CMOS technology. *IEEE Journal of Solid-State Circuits*, 50(1):150–157, 2015.

59. Ruth Brain, A Baran, N Bisnik, H-P Chen, S-J Choi, A Chugh, M Fradkin, T Glassman, F Hamzaoglu, E Hoggan, et al. A 22nm high performance embedded DRAM SoC technology featuring tri-gate transistors and MIMCAP COB. In *VLSI Circuits (VLSIC), 2013 Symposium on*, pages T16–T17. IEEE, 2013.

60. Yih Wang, Umut Arslan, Nabhendra Bisnik, Ruth Brain, Swaroop Ghosh, Fatih Hamzaoglu, Nick Lindert, Mesut Meterelliyoz, Joodong Park, Shigeki Tomishima, et al. Retention time optimization for eDRAM in 22nm tri-gate CMOS technology. In *Electron Devices Meeting (IEDM), 2013 IEEE International*, pages 9–5. IEEE, 2013.

61. JEDEC. JEDEC DDR4 SDRAM STANDARD (JEDS79-4). http://www.jdec.org/

62. Arnab Biswas and Adrian M Ionescu. 1T capacitor-less DRAM cell based on asymmetric tunnel FET design. *IEEE Journal of the Electron Devices Society*, 3(3):217–222, 2015.

63. Navneet Gupta, Adam Makosiej, Andrei Vladimirescu, Amara Amara, and Costin Anghel. Tunnel FET based refresh-free-DRAM. In *Proceedings of the Conference on Design, Automation & Test in Europe*, pages 914–917. European Design and Automation Association, 2017.

64. Navneet Gupta, Adam Makosiej, Andrei Vladimirescu, Amara Amara, and Costin Anghel. Tunnel FET based ultra-low-leakage compact 2T1C SRAM. In *Quality Electronic Design (ISQED), 2017 18th International Symposium on*, pages 71–75. IEEE, 2017.
65. Ki-Whan Song, Jin-Young Kim, Jae-Man Yoon, Sua Kim, Huijung Kim, Hyun-Woo Chung, Hyungi Kim, Kanguk Kim, Hwan-Wook Park, Hyun Chul Kang, et al. A 31 ns Random Cycle VCAT-Based $4F^2$ DRAM With Manufacturability and Enhanced Cell Efficiency. *IEEE Journal of Solid-State Circuits*, 45(4):880–888, 2010.
66. Navneet Gupta, Adam Makosiej, Andrei Vladimirescu, Amara Amara, and Costin Anghel. 1.56 GHz/0.9 V energy-efficient reconfigurable CAM/SRAM using 6T-CMOS bitcell. In *ESSCIRC 2017-43rd IEEE European Solid State Circuits Conference*, pages 316–319. IEEE, 2017.
67. Dejan Markovic, Borivoje Nikolic, and Robert Brodersen. Analysis and design of low-energy flip-flops. In *Proceedings of the 2001 international symposium on Low power electronics and design*, pages 52–55. ACM, 2001.
68. R Ramanarayanan, N Vijaykrishnan, and MJ Irwin. Characterizing dynamic and leakage power behavior in flip-flops. In *ASIC/SOC Conference, 2002. 15th Annual IEEE International*, pages 433–437. IEEE, 2002.
69. SH Rasouli, A Amirabadi, A Seyedi, and Ali Afzali-Kusha. Double edge triggered feedback flip-flop in sub 100nm technology. In *Design Automation, 2006. Asia and South Pacific Conference on*, pages 6–pp. IEEE, 2006.
70. Matthew Cotter, Huichu Liu, Suman Datta, and Vijaykrishnan Narayanan. Evaluation of tunnel FET-based flip-flop designs for low power, high performance applications. In *Quality electronic design (ISQED), 2013 14th international symposium on*, pages 430–437. IEEE, 2013.
71. Navneet Gupta, Adam Makosiej, Andrei Vladimirescu, Amara Amara, and Costin Anghel. Ultra-low-power compact TFET flip-flop design for high-performance low-voltage applications. In *Quality Electronic Design (ISQED), 2016 17th International Symposium on*, pages 107–112. IEEE, 2016.
72. Greg Burda, Yesh Kolla, Jim Dieffenderfer, and Fadi Hamdan. A 45nm CMOS 13-port 64-word 41b fully associative content-addressable register file. In *Solid-State Circuits Conference Digest of Technical Papers (ISSCC), 2010 IEEE International*, pages 286–287. IEEE, 2010.
73. Xilinx. XA Zynq-7000 All Programmable SoC First Generation Architecture. www.xilinx.com
74. Jishen Zhao, Cong Xu, and Yuan Xie. Bandwidth-aware reconfigurable cache design with hybrid memory technologies. In *Proceedings of the International Conference on Computer-Aided Design*, pages 48–55. IEEE Press, 2011.
75. AD Santana Gil, FJ Quiles Latorre, M Hernandez Calvino, E Herruzo Gomez, and JI Benavides Benitez. Optimizing the physical implementation of a reconfigurable cache. In *Reconfigurable Computing and FPGAs (ReConFig), 2012 International Conference on*, pages 1–6. IEEE, 2012.
76. George Kalokerinos, Vassilis Papaefstathiou, George Nikiforos, Stamatis Kavadias, Manolis Katevenis, Dionisios Pnevmatikatos, and Xiaojun Yang. FPGA implementation of a configurable cache/scratchpad memory with virtualized user-level RDMA capability. In *Systems, Architectures, Modeling, and Simulation, 2009. SAMOS'09. International Symposium on*, pages 149–156. IEEE, 2009.
77. Yen-Jen Chang, Kun-Lin Tsai, and Hsiang-Jen Tsai. Low leakage TCAM for IP lookup using two-side self-gating. *IEEE Transactions on Circuits and Systems I: Regular Papers*, 60(6):1478–1486, 2013.
78. S Matsunaga, N Sakimura, R Nebashi, Y Tsuji, A Morioka, T Sugibayashi, S Miura, H Honjo, K Kinoshita, H Sato, et al. Fabrication of a 99%-energy-less nonvolatile multi-functional CAM chip using hierarchical power gating for a massively-parallel full-text-search engine. In *VLSI Circuits (VLSIC), 2013 Symposium on*, pages C106–C107. IEEE, 2013.

79. Navneet Gupta, Adam Makosiej, Andrei Vladimirescu, Amara Amara, and Costin Anghel. 16Kb hybrid TFET/CMOS reconfigurable CAM/SRAM array based on 9T-TFET bitcell. In *Solid-State Device Research Conference (ESSDERC), 2016 46th European*, pages 356–359. IEEE, 2016.

80. Content addressable memory (CAM) devices having dedicated mask cell sub-arrays therein and methods of operating same, author=Proebsting, Robert J and Park, Kee and Chu, Scott Yu-Fan, January 4 2005. US Patent 6,839,256.

81. Amit Agarwal, Steven Hsu, Sanu Mathew, Mark Anders, Himanshu Kaul, Farhana Sheikh, and Ram Krishnamurthy. A 128× 128b high-speed wide-and match-line content addressable memory in 32nm CMOS. In *ESSCIRC (ESSCIRC), 2011 Proceedings of the*, pages 83–86. IEEE, 2011.

82. Anh Tuan Do, Chun Yin, Kiat Seng Yeo, and Tony Tae-Hyoung Kim. Design of a power-efficient CAM using automated background checking scheme for small match line swing. In *ESSCIRC (ESSCIRC), 2013 Proceedings of the*, pages 209–212. IEEE, 2013.

83. Chua-Chin Wang, Chia-Hao Hsu, Chi-Chun Huang, and Jun-Han Wu. A self-disabled sensing technique for content-addressable memories. *IEEE Transactions on Circuits and Systems II: Express Briefs*, 57(1):31–35, 2010.

84. Mrigank Sharad, Deliang Fan, and Kaushik Roy. Ultra low power associative computing with spin neurons and resistive crossbar memory. In *Design Automation Conference (DAC), 2013 50th ACM/EDAC/IEEE*, pages 1–6. IEEE, 2013.

85. Yuanfan Yang, Jimson Mathew, and Dhiraj K Pradhan. Matching in memristor based auto-associative memory with application to pattern recognition. In *Signal Processing (ICSP), 2014 12th International Conference on*, pages 1463–1468. IEEE, 2014.

86. Amirali Ghofrani, Abbas Rahimi, Miguel A Lastras-Montaño, Luca Benini, Rajesh K Gupta, and Kwang-Ting Cheng. Associative Memristive Memory for Approximate Computing in GPUs. *IEEE Journal on Emerging and Selected Topics in Circuits and Systems*, 6(2):222–234, 2016.

87. Mohsen Imani, Abbas Rahimi, and Tajana S Rosing. Resistive configurable associative memory for approximate computing. In *Design, Automation & Test in Europe Conference & Exhibition (DATE), 2016*, pages 1327–1332. IEEE, 2016.

88. Mohsen Imani, Pietro Mercati, and Tajana Rosing. ReMAM: low-energy resistive multi-stage associative memory for energy efficient computing. In *Quality Electronic Design (ISQED), 2016 17th International Symposium on*, pages 101–106. IEEE, 2016.

89. Mohsen Imani, Shruti Patil, and Tajana Rosing. Approximate computing using multiple-access single-charge associative memory. *IEEE Transactions on Emerging Topics in Computing*, 2016.

90. Hang Zhang, Mateja Putic, and John Lach. Low-power gpgpu computation with imprecise hardware. In *Proceedings of the 51st Annual Design Automation Conference*, pages 1–6. ACM, 2014.

91. Syed Z Gilani, Nam Sung Kim, and Michael Schulte. Scratchpad memory optimizations for digital signal processing applications. In *Design, Automation & Test in Europe Conference & Exhibition (DATE), 2011*, pages 1–6. IEEE, 2011.

92. Fabio Frustaci, Mahmood Khayatzadeh, David Blaauw, Dennis Sylvester, and Massimo Alioto. A 32kb SRAM for error-free and error-tolerant applications with dynamic energy-quality management in 28nm CMOS. In *Solid-State Circuits Conference Digest of Technical Papers (ISSCC), 2014 IEEE International*, pages 244–245. IEEE, 2014.

93. Navneet Gupta, Prashant Dubey, Shaileshkumar Pathak, Kaushik Saha, Ashish Kumar, and R Sai KRISHNA. Memory architecture and design methodology with adaptive read, May 27 2014. US Patent 8,737,144.

94. Daeyeon Kim, Vikas Chandra, Robert Aitken, David Blaauw, and Dennis Sylvester. An adaptive write word-line pulse width and voltage modulation architecture for bit-interleaved 8T SRAMs. In *Proceedings of the 2012 ACM/IEEE international symposium on Low power electronics and design*, pages 91–96. ACM, 2012.

95. N Gupta, A Makosiej, O Thomas, A Amara, A Vladimirescu, and C Anghel. TFET/CMOS hybrid pseudo dual-port SRAM for scratchpad applications. In *Ultimate Integration on Silicon (EUROSOI-ULIS), 2015 Joint International EUROSOI Workshop and International Conference on*, pages 209–212. IEEE, 2015.
96. Navneet Gupta, Adam Makosiej, Andrei Vladimirescu, Amara Amara, Sorin Cotofana, and Costin Anghel. TFET NDR skewed inverter based sensing method. In *Nanoscale Architectures (NANOARCH), 2016 IEEE/ACM International Symposium on*, pages 13–14. IEEE, 2016.

Index

A

Adaptive memories, 105, 106, 114, 119, 121
Adaptive *read* technique, 105
  asynchronous memory design, 119–120
  AVS, 113
  bitline discharge, 113
  differential voltage (Vdiff), 113
  functionality
    operation RdOK, 115
    RdOK signal generation, 116
    read operations, 114
    self-adapting SRAM using RdOK,
      117–119
    SRAM with self-tuning using RdOK,
      116–117
  operation, SA, 113–114
  proposed SA, 114
Advance voltage scaling (AVS), 113
Ambipolar effect, 6–8
Approximate-search, 3, 74, 100
Array-based 6T TFET SRAM
  3 × 3 bitcell array, 20
  corner cells RET, 19
  HS problem, 19–20
  nine-cell fragment, 19
  WD problem, 20–21
Associative memories
  with CMOS CAM bitcell
    associative search operation, 92–94
    optimized comparison logic, 90
    6T-CAM bitcell, 90, 91
    two-bit WTA-based comparison logic,
      91, 92
    write operation, 94–95
  current-mode WTA logic, 97–100

8T-TFET associative memory bitcell, 74,
  95–97
  neuromorphic implementations, 90
  read technique, 119
  ReRAMs, 90
  search applications, 90
  standard 10T-NOR CAM bitcell, 97
  STTRAMs, 90

B

Band to band tunneling (BTBT), 5–7, 10
Battery-powered devices, 124
Binary CAM (BCAM), 83–85
Built-in-self-test (BIST), 106, 116

C

Charge-injection sense amplifier
  MOS capacitors, 84
  MOSFET capacitors, 107
  single-ended imbalanced SA, 30, 106–108
    with open-bitline, 108–110
CMOS reconfigurable CAM/SRAM
  comparison, 88, 89
  design, 82
  high-speed 6T-CMOS, 88
  match line, 82
  read/write/search test patterns, 86
  ReCSAM architecture
    BCAM, 83–85
    imbalance tuning (IT), 86–87
    memory organization, 83
    pseudo-TCAM mode, 86
    single-bitline configuration, 83

© Springer Nature Switzerland AG 2021
N. Gupta et al., *TFET Integrated Circuits*,
https://doi.org/10.1007/978-3-030-55119-3

CMOS reconfigurable CAM/SRAM (*cont.*)
    speed improvements, 88
    SRAM mode, 86
    search speed, 88, 90
Complementary-metal-oxide-semiconductor
        (CMOS)
    conventional differential sensing, 105
    FDSOI, 3–4
    flip-flops, 59, 60 (*see also* Flip-flops,
        TFET)
    hybrid TFET-CMOS multi-core processor,
        124
    memory circuits, 123
    performance improvements, 1
    ReCSAM (*see* Reconfigurable
        CAM/SRAM (ReCSAM))
    vs. TFET, 3, 6
        dynamic and static power *vs.* supply
            voltage, 14
        dynamic performance, 11
        Miller capacitance effect, 13
        operating frequency, 12
        PDP, 12, 13
    traditional reliance, 2
Content-addressable memories (CAMs), 3
    approximate-search, 74, 100
    in ASIC, 73, 90
    associative memory architecture (*see*
        Associative memories)
    as cache memory, 73
    to CMOS on 6T-CMOS bitcell (*see* CMOS
        reconfigurable CAM/SRAM)
    flexible architectures, 73
    FPGA-based, 73
    9T-TFET CAM cell (*see* Reconfigurable
        CAM/SRAM (ReCSAM))
    ultra-low-power (*see* Ultra-low-power
        ternary CAM)
Conventional differential sensing, 105
Current-mode WTA circuit, 95, 97, 99

D
Dual data rate (DDR), 46, 52, 55, 57
Dual-port associative memory, 95–97
Dual-port SRAM (DPSRAM)
    DVFS, 26
    8T-TFET bitcell, 17
    8T-TFET dual-port SRAM cell, 28
    in embedded systems, 26
    32 Kb Mux-8 dual-port memory
        architecture, 29
    micro-architecture design, 29
    and multi-port, 26

open bitlines architecture, 108, 109
performance and stability, 8T DPSRAM,
        32–33
proposed TFET cell, 27–29
pseudo dual-port scratchpad, 27
read operation, 29–30
single-ended sensing, 30
write assist, 26
write operation, 30–32
Dynamic power, 9, 11, 13–15, 65, 66, 72, 82,
        105, 125
Dynamic random access memory (DRAM), 2,
        3
    capacitor-less TFET, 46
    design challenges, 46
    embedded (*see* Embedded DRAMs
        (eDRAMs))
    and Flash, 124
    ultimate (*see* Ultimate DRAM (uDRAM))
Dynamic voltage frequency scaling (DVFS),
        26, 32
    EDP-and barrier-aware, 124–125
    IoT, 2

E
Effective oxide thickness (EOT), 45, 48, 51, 52
Embedded DRAMs (eDRAMs)
    array density, 45
    capacitor size, 45–46
    and conventional DRAMs, 51
    density, 45
    EOT, 45
    and LVSRAM, 57
    NWL, 45
    planar process, 46
    2T1C uSRAM, 52
Embedded Flash (eFlash), 45
Energy-delay product (EDP), 124
Energy-efficient systems, 1

F
Flip-flops, TFET
    CMOS, 59
    design-level issues, 60
    FinFET, 64–72
    limitations, 60
    master-slave, 61
    with modified latch, 61, 62
    MOSFET master-slave pseudo-static D
        design, 62
    MSFF (*see* Master-slave flip-flop (MSFF))
    pseudo-static flip-flop, 63

semi-dynamic design, 61
sense amplifier-based, 61, 62
speed comparison, 62
synchronous logic and microprocessor-
          based systems, 59
transmission-gate, 60, 61, 68
Fully depleted silicon on insulator (FDSOI),
          3–5, 29, 39, 46, 53, 60, 74, 83, 86,
          100, 120

G
Gate capacitance, 10, 11

H
Half-selection (*HS*), 17, 19–20, 25, 30, 32, 76,
          77, 93, 95
Heterojunction TFET (HTFET), 5, 13, 21, 61,
          70, 123, 127
Hold time, 67
Homojunction TFET, 5, 61

I
Imbalanced sense amplifier
    charge-injection based single-ended SA,
          30, 106–108
      with open-bitline, 108–110
    MOS capacitors, 85
    read speed with/without SA imbalance
          tuning, 87
International Technology Roadmap for
          Semiconductors (ITRS), 3
Internet-of-Things (IoT)
    cost and power efficiency, 59
    DVFS, 2
    energy-efficient systems, 1
    memory power, optimization, 2
    ultra-low power consumption, 1
    and WSN, 1

L
Leakage power, 26, 33, 42, 65–67, 70, 74, 75
Low power
    CAM-based, 90
    dynamic power efficiency, 14
    flip-flop designs, 59
    LSTP, 3, 15, 17
    PDP analysis, 12
    SRAM-based, 95
    TFETs, 2, 3
    3T-TFET SRAM cell, 34

ultra-low power TFET TCAM, 100–103
Low-standby power (LSTP), 3, 15, 17, 125
    FinFET-LSTP designs, 65–68
    reconfigurability, 74

M
Master-slave flip-flop (MSFF)
    TFET-FF design, 63
    12-TFETs
        energy efficiency, 65–66
        flip-flop operation, 64–65
        flip-flop performance, 66–68
        proposed TFET-FF design, 63, 64
Multi-port SRAMs, 26

N
Negative Differential Resistance (NDR), 3, 7,
          23, 46, 55, 60, 100, 125
        See also TFET *NDR* based sense
          amplifier (SA)
Negative wordline (NWL), 42, 45, 46
Non-volatile devices, 124
Non-volatile-memory (NVM), 90, 105
N-type (NTFET), 5, 6, 10–12, 22

O
On-chip storage demand, 45

P
Pattern search, 74, 100
Power delay product (PDP), 12
Predictive Technology Model (PTM), 8,
          10–12, 14
Pseudo Ternary CAM (TCAM), 74, 82, 86, 89,
          100–103
P-type (PTFET), 5, 6, 9–12, 15

R
Read assist (*RA*), 79, 80, 84, 92, 93, 116
Reconfigurable CAM/SRAM (ReCSAM)
    CMOS bitcell, 74, 82 (*see also* CMOS
          reconfigurable CAM/SRAM)
    memory array, CAM and SRAM, 74
    9T-TFET CAM cell
        dual wordline, 75
        performance, 79–80
        power consumption, 80–82
        proposed design, 75, 76
        read operation, 75, 77–79

Reconfigurable CAM/SRAM
    (ReCSAM) (*cont.*)
    write operation, 76–77
    TFET/CMOS hybrid, 74
    ultra-low leakage, 82
Refresh-free DRAM, 46, 52, 55
Ring oscillator (RO), 11–14

S
Self-tuning memories, 116, 118
Sense amplifier (SA)
    adaptive *read* (*see* Adaptive *read* technique)
    balanced single-ended, 106
    DPSRAM memory, 108, 109
    imbalanced, 106, 107
    imbalanced SA with open-bitline,
        108–110
    inverter-based, 105
    read scheme, 106
Setup time (T_{Setup}), 64, 66–70
7T-TFET ternary CAM bitcell, *see* Ultra-low-
        power ternary CAM
Silicon-TFET
    CMOS memory circuits, 123
    device structure, 8
    low-k spacer and high-k gate dielectrics, 8,
        9
    NTFET, 10
    PTFET, 9
    reverse biasing, 11
    Si-CMOS, 13
    SoC, 124
    and SPICE model, 11
    TCAD simulations, 8
    3T-TFET bitcell, 34
    turn-on voltage, 8
    ULP applications, 59, 60
    unidirectional, 10
    variations and modelling, 9
Single-ended sensing, 37, 40
    charge injection based imbalanced SA,
        106–108
    conventional differential sensing, 105
    differential SA, 105
    DPSRAM, 30
    imbalanced SA, 84, 85
    imbalanced SA with open-bitline, 108–110
    inverter-base sensing, 30
    optimized TFET memory cells, 105
    SA and adaptive memory, 119–120
    TFET *NDR* skewed inverter-based method,
        110–113
6T-CMOS CAM bitcell, 74, 91

Skewed inverter based sense amplifier,
        110–113
SPICE TFET model, 11
Static noise margin (SNM), 19–21, 26, 33, 36
Static power, 2, 3, 14, 65, 71
Static Random Access Memory (SRAM)
    adaptive techniques, 105
    conventional ECC, 25
    hetero-junction TFET, 21
    modified 6T TFET bitcell, 21, 23
    multi-port SRAMs, 26
    NDR property, 23
    on-chip, 45
    standard 6T-SRAM-bitcell based design,
        124
    2TIC uSRAM, 52–54
    3T-TFET bitcells (*see* TFET-CMOS hybrid
        SRAM)
    4T-*NDR* TFET bitcells, 24
    5T-TFET-CMOS hybrid bitcell, 23
    6T-TFET operation
        array-based analysis, 19–21
        read and *write* mechanisms, 18
        stability, 19
    7T-TFET bitcells, 21, 22
    8T CMOS SRAM, 25
    8T-TFET bitcells, 21, 22
    8T-TFET dual-wordline bitcell, 26
    uDRAM, 46
System-on-chip (SoC), 1, 45, 59, 73, 74, 124

T
TCAD device, 3, 8, 9, 11
Technology scaling, 1
TFET-CMOS co-integration, 124, 125
TFET-CMOS hybrid memories, 17, 106, 120
TFET-CMOS hybrid SRAM, 17
    dynamic power consumption, 34
    energy efficiency, 40–42
    memory architecture, 38–39
    memory layout, 39–40
    NDR property, 34
    proposed cell architectures, 34, 35
    read and *write* performance (WLPcrit),
        42–43
    read operation, 35, 37, 38
    retention mode and stability, 35–36
    write operation, 36–37
TFET *NDR* based sense amplifier (SA)
    DC characteristics, 111
    MOSFETs, 112
    read scalability, circuit, 111, 112
    sense amplifier schematic, 111

sensing method, 111–112
TFET-CMOS circuit, 112
trip points, 112
ultra-compact skewed inverter-based
     sensing method, 110
Transmission gate flip-flop, 60, 61, 68
Tri-state TFET inverter, 61, 63, 64, 71, 110
Tri-state TFET latch, 63, 64, 71
Tunnel-field-effect-transistors (TFET)
     ambipolar effect, 6, 7
     battery-powered applications, 124
     vs. CMOS, 3, 6
     device symbols, 15
     EDP-and barrier-aware DVFS, 124–125
     flip-flops (see Flip-flops, TFET)
     homo-and hetero-junction, 5
     I_D (V_{GS}, V_{DS}) characteristics, 6–11
     low-power circuits and systems
          applications, 2
     in LSTP applications, 17
     memory bitcells, 105
     and MOSFETs, 5, 7, 46
     operation, 5
     p-i-n gated junctions, 5
     potential, 3
     quantum-tunneling effect, 2
     ReCSAM (see Reconfigurable
          CAM/SRAM (ReCSAM))
     in reverse bias, 6, 8
     TFET-CMOS hybrid cores, 125
     tri-state inverters, 61
12T-TFET master-slave flip-flop (MSFF)
     comparison
          clock-to-output propagation delay
               (T_{CP2Q}), 69–70
          leakage, 70–71
          setup time (T_{Setup}), 66–70
          $T_{Critical}$, 70
          transistor count, 68
     energy efficiency, 65–66
     flip-flop operation, 64–65
     flip-flop performance, 66–68
     proposed TFET-FF design, 63, 64

U
Ultimate CAM (uCAM), 54–57
Ultimate DRAM (uDRAM)
     advantages, 47

NDR, 46
retention, 47
static latch behavior, 46
1T1C uDRAM cell
     advantages, 47
     bitcell implementation and
          performance, 51–52
     memory array organization, 51
     NDR, 46
     read operation, 49–51
     retention, 47
     static latch behavior, 46
     TFET DRAM bitcell, 47
     write operation, 48–49
2T1C uSRAM bitcell
     bitcell layout, 54
     implementation, 52
     memory array organization, 53
     RBL and WBL, 52–53
3T1C CAM bitcell, 54–55
TFET DRAM bitcell, 47
write operation, 48–49
Ultimate SRAM (uSRAM), 52–54, 56, 57
Ultra-low power, 1, 34, 100–104
Ultra-low power memories, 17
Ultra-low-power ternary CAM
     hybrid TFET/CMOS 7T-TCAM bitcell,
          100, 101, 103
     memory array organization, 102
     multi-bit latch, 101–102
     read operation, 101, 102
     TCAM cell, 100–101
     TCAM dual-bitcell layout, 103
     write operation, 101, 102

W
Winner-take-all (WTA) logic, 90–99
Wireless-Sensor Nodes (WSN)
     cost and power efficiency, 59
     energy-efficient systems, 1
     and IoT, 1
     ultra-low power consumption, 1
WordLine Pulse width, 29–32, 42, 43
Write assist (WA), 26, 30–32, 43, 79, 80,
     106
Write-disturb (WD), 17, 20–22, 25, 28, 30, 32,
     79
Write wordline (WWL), 101, 102, 106

Printed in the United States
by Baker & Taylor Publisher Services